# Collector's Guide to the
# PYROXENE GROUP

*Schiffer Earth Science Monographs Volume 7*

Robert J. Lauf

4880 Lower Valley Road, Atglen, Pennsylvania 19310

# Dedication

This book is dedicated to Jim DeBruzzi, a gifted science teacher and a truly inspiring mentor.

**Other Schiffer Books by Robert J. Lauf**
*Collector's Guide to the Axinite Group.*
ISBN: 9780764332166. $19.99
*Collector's Guide to the Epidote Group.*
ISBN: 9780764330483. $19.99
*Collector's Guide to the Mica Group.*
ISBN: 9780764330476. $19.99
*Collector's Guide to the Three Phases of Titania: Rutile, Anatase, and Brookite.* ISBN: 9780764332685. $19.99
*Collector's Guide to the Vesuvianite Group.*
ISBN: 9780764332159. $19.99
*Introduction to Radioactive Minerals.*
ISBN: 9780764329128. $29.95

**Other Schiffer Books on Related Subjects**
*Collecting Fluorescent Minerals.* Stuart Schneider.
ISBN: 0764320912. $29.95
*Collector's Guide to Fluorite.* Arvid Eric Pasto. ISBN: 9780764331930. $19.99
*The World of Fluorescent Minerals.* Stuart Schneider.
ISBN: 0764325442. $29.95

Copyright © 2010 by Robert J. Lauf
Library of Congress Control Number: 2009936480

All rights reserved. No part of this work may be reproduced or used in any form or by any means—graphic, electronic, or mechanical, including photocopying or information storage and retrieval systems—without written permission from the publisher.
The scanning, uploading and distribution of this book or any part thereof via the Internet or via any other means without the permission of the publisher is illegal and punishable by law. Please purchase only authorized editions and do not participate in or encourage the electronic piracy of copyrighted materials.
"Schiffer," "Schiffer Publishing Ltd. & Design," and the "Design of pen and ink well" are registered trademarks of Schiffer Publishing Ltd.

Designed by Mark David Bowyer
Type set in Arno Pro / Humanist521 BT

ISBN: 978-0-7643-3404-7
Printed in China

Schiffer Books are available at special discounts for bulk purchases for sales promotions or premiums. Special editions, including personalized covers, corporate imprints, and excerpts can be created in large quantities for special needs. For more information contact the publisher:

Published by Schiffer Publishing Ltd.
4880 Lower Valley Road
Atglen, PA 19310
Phone: (610) 593-1777; Fax: (610) 593-2002
E-mail: Info@schifferbooks.com

For the largest selection of fine reference books on this and related subjects, please visit our web site at
**www.schifferbooks.com**
We are always looking for people to write books on new and related subjects. If you have an idea for a book please contact us at the above address.

This book may be purchased from the publisher.
Include $5.00 for shipping.
Please try your bookstore first.
You may write for a free catalog.

In Europe, Schiffer books are distributed by
Bushwood Books
6 Marksbury Ave.
Kew Gardens
Surrey TW9 4JF England
Phone: 44 (0) 20 8392 8585; Fax: 44 (0) 20 8392 9876
E-mail: info@bushwoodbooks.co.uk
Website: www.bushwoodbooks.co.uk

# Contents

**Preface** ...................................................... 5

**Acknowledgments** .................................... 7

**Introduction** ............................................... 8

**Taxonomy of the Pyroxene Group** ........... 14
   General Formula and Subdivisions of the Group ......... 14
      Mg-Fe pyroxenes ............................................. 14
      Mn-Mg pyroxenes ............................................ 14
      Ca pyroxenes .................................................. 14
      Ca-Na pyroxenes ............................................ 14
      Na pyroxenes ................................................. 14
      Li pyroxenes .................................................. 14
   Crystal Structure and Morphology ........................ 18
   Crystal Chemistry ................................................ 22

**Formation and Geochemistry** .................. 23
   Pyroxenes in Igneous Rocks ................................ 23
   Pyroxenes in Metamorphic Rocks ....................... 25
   Pyroxenes in Extraterrestrial Rocks ..................... 31

**The Minerals** ............................................ 37
   Aegirine ............................................................. 37
   Aegirine-augite .................................................. 43
   Augite ................................................................ 44
   Clinoenstatite ..................................................... 46
   Clinoferrosilite ................................................... 46

- Diopside ...... 47
- Donpeacorite ...... 64
- Enstatite ...... 65
- Esseneite ...... 68
- Ferrosilite ...... 69
- Hedenbergite ...... 70
- Jadeite ...... 73
- Jervisite ...... 75
- Johannsenite ...... 75
- Kanoite ...... 77
- Kosmochlor ...... 78
- Namansilite ...... 80
- Natalyite ...... 81
- Omphacite ...... 81
- Petedunnite ...... 83
- Pigeonite ...... 84
- Spodumene ...... 85

**References** ...... 89

# Preface

This volume continues a series of monographs on important groups of so-called rock forming silicates, the purpose of which is to help mineral collectors gain a better appreciation of these complex minerals. Because of the importance of rock forming minerals in geological processes, they are the subject of extensive published research, much of which has been brought together in the five-volume compendium *Rock-Forming Minerals* (Deer, Howie, and Zussman 1962) and the greatly expanded Second Edition thereof. Among rock-forming minerals, the pyroxene group is perhaps best known to collectors through well-formed diopside crystals long known from many localities, including those in Canada, the northeastern United States, and the Alpine regions of Europe, and the gem-quality spodumene found in pegmatites from California to Pakistan. Spectacular recent finds of diopside in Pakistan, hedenbergite at Dal'negorsk, Russia, and aegirine in Malawi have added to the interest in this mineral group. Several pyroxene species are important in the gem trade, including colored varieties of spodumene (*kunzite* and *hiddenite*), green "chrome" diopside, and jadeite, one of the two true jade minerals. The present monograph is organized as follows: After a brief introduction, the general treatment begins with an explanation of the chemistry and taxonomy of the group and a discussion of ongoing research. A section on their formation and geochemistry explains the kinds of environments where pyroxenes are formed. Then, a detailed entry for each mineral provides information on important localities and full-color photos wherever possible so that collectors can see what good specimens look like and which minerals one might expect to find in association with pyroxenes. An extensive bibliography is provided for those readers who wish to learn more about particular topics.

As in earlier volumes in this series, the photographs were not selected to showcase extremely expensive "museum pieces" or purported "best in the world" specimens, but instead, representative examples that illustrate the rich variety of colors and habits in specimens that an interested collector could actually hope to obtain and study. All specimens are from the author's research collection.

Currently-accepted nomenclature (Morimoto et al. 1989) is used for the individual pyroxene species, but obsolete and varietal names are also provided to help the reader interpret earlier literature on the group.

# Acknowledgments

The following colleagues kindly provided technical information, literature, and helpful discussions: Dr. Larry Anovitz, *Oak Ridge National Laboratory*; Deborah Cole, *Oak Ridge National Laboratory*; Dr. Arvid Pasto. Important specimens and background information were supplied by: John Betts; Dudley Blauwet, *Mountain Minerals*; Dave Bunk; Sharon Cisneros, *Mineralogical Research Co.*; John Cleary, *Ventana Mining*; Richard Dale, *Dale Minerals*; Gunnar Färber; Shields Flynn, *Trafford-Flynn Minerals*; Robert Haag, *Robert Haag Meteorites*; Pete Heckscher, *The Crystal Circle*; Leonard Himes, *Minerals America*; Mohammed Javed, *Javed's International Gem Imports*; Rob Kulakofski, *Color-Wright*; Rob Lavinsky, *The Arkenstone*; Marlene Leitzman; Janice Muller, *New Era Gems*; Tony Nikischer, *Excalibur Mineral Co.*; Neal Pfaff, *M. Phantom Minerals*; C. Carter Rich; Jeff Schlottman, *Crystal Perfection*; Jaye Smith, *The Rocksmiths*; Scott Wallace, *Majestic Minerals*; and Chris Wright, *Wright's Rock Shop*.

# Introduction

The pyroxene group comprises some twenty recognized species, of which a few are fairly common and others are quite rare. As a microscopic constituent of basalt, pyroxene is widespread in the earth's crust; in fact, as noted by Deer, Howie, and Zussman (1978), "Pyroxenes are the most important group of ferromagnesian silicates, and occur as stable phases in almost every type of igneous rock. They are found also in many rocks of widely different compositions formed under conditions of both regional and thermal metamorphism." The group has been the subject of a tremendous amount of research, driven in part by studies of lunar samples. According to Prewitt (1980), "The influence of the lunar program on pyroxene research was extraordinary, and our understanding of pyroxene relationships in terrestrial occurrences benefited tremendously because the lunar pyroxenes provided a basis for comparison with the more complex chemical and structural behavior of terrestrial environments."

Among the species of this group, a half-dozen or more are commonly found in well-developed crystals that are of most interest to mineral collectors. These include aegirine, augite, diopside, enstatite, hedenbergite, johannsenite, and spodumene. For the rare species collector, a quick survey indicates that practically all of the pyroxenes seem to be readily available from various rare mineral dealers.

Spodumene, as one of the principal ores of lithium, is a significant industrial commodity. It is used as a feedstock for the production of lithium carbonate and lithium metal, and is also used directly in making various glass and ceramic materials, and in lithium batteries. In glass manufacturing, lithium compounds act as a powerful flux to reduce the melting temperature and lower fuel costs, but lithium also modifies the glass properties in several important ways. It is especially useful in lowering the viscosity of molten glass in order to run glassmaking operations faster and to mold more intricate articles. The development of pyroceramic cookware is based on the formation of the ß-spodumene phase that reduces the glass-ceramic's thermal expansion coefficient practically to zero. According to recent figures from the U.S. Geological Survey, in 2005 the total worldwide lithium market averaged 15,000 metric tons of Li contained in minerals and compounds. Consumption of Li for secondary batteries is growing at a much higher rate than for other applications, representing 9% of the market in 2000 and 20% in 2005. The recent interest in electric vehicles is likely to create significantly more growth in the Li market over the next few decades. Roughly 60% of the worldwide supply of Li minerals was produced at the Greenbushes spodumene mine in Australia. A spodumene mine at Bernic Lake, Manitoba, Canada, has operated on a commercial scale since 1986. China produces large quantities of lithium carbonate from spodumene ore, both domestically mined and imported from Australia.

Several pyroxenes are important in the gem and lapidary arts. These include diopside, enstatite (including its varieties *bronzite* and *hypersthene*), spodumene (especially its varieties *kunzite* and *hiddenite*), and of course jadeite, which is one of the true jade minerals.

Massive forms of enstatite from several locales make interesting cabbing materials. Enstatite var. *bronzite* from Brazil contains randomly oriented grains having a fibrous texture. The bundles of acicular crystallites, in subdued shades of reddish to golden brown, catch the light in an attractive way similarly to tiger-eye. A dark form of enstatite var. *hypersthene* from Canada has an exsolution texture that creates a subtle herringbone pattern.

Figure 1. Slabs of enstatite var. *bronzite* from Brazil showing a silky, textured patchwork of reds and browns somewhat like tiger-eye. The slabs were moistened to bring out the color and photographed using a flatbed scanner. *RJL3468*

Figure 2. Freeform cabochon of enstatite var. *bronzite* from Brazil, showing a subtle but luminous mixture of reds and golds.

Figure 3. Teardrop-shaped cabochon about 30 mm tall, of enstatite var. *hypersthene* from Sagueway, Quebec, Canada, showing reflections from the fine lamellar exsolution structure.

# 10 Introduction

Transparent, faceting-grade enstatite in shades of green to brown is found at several localities, including Myanmar (Burma), Sri Lanka, Tanzania, Norway, Germany, Arizona, and California. Star enstatite is found in India; the stones show a four-rayed star. Gem-grade enstatite var. *hypersthene* is found in Baja, California, Mexico (Arem 1977). Green to gray cat's-eye enstatite is found in Sri Lanka. *Bastite* is a decomposition product of enstatite to serpentine; it has a silky sheen and is sometimes used for cabochons.

Figure 5. A simple gold pendant set with a 3-carat round enstatite; the rough stone was mined in Namanga, Kenya.

Figure 4. Rough pieces of faceting-grade enstatite var. *bronzite* from Tanzania, totaling about 60 carats, and a 5.6-carat oval enstatite from Namanga, Kenya.

Gem-grade diopside is occasionally faceted. The stones are usually some shade of green and those with a deeper green color are sometimes marketed as "chrome diopside" even when the color is more likely due to iron. Bright emerald-green diopside is found in the Merelani Hills, Tanzania; its color is attributed to chromium and vanadium. Star diopside is an interesting phenomenal stone found in Nammakal, in the province of Tamilnadu, southern India. Properly oriented cabochons display a four-rayed star, and have long been used as a low-cost substitute for black star sapphire. The asterism is attributed to microscopic inclusions of magnetite and tremolite aligned parallel to the [001] and [100] planes in the diopside crystal (Doukhan et al. 1990).

Figure 6. A brilliant green thumbnail-sized crystal of diopside from the Merelani Hills, Tanzania, with a small amount of black crystalline graphite attached. *RJL3061*.

Figure 7. Two faceted diopsides: a 2.6-carat oval-cut "chrome" diopside from Sri Lanka (left) and a darker green 1.5-carat round stone from Tanzania (right).

Figure 8. A 63-carat black star diopside cabochon from Nammakal, India, showing the characteristic four-rayed star that arises from light scattering off tiny oriented inclusions of magnetite and tremolite. RJL3519

Figure 9. Photomicrograph of the star diopside cabochon from the previous figure, showing the plate-like inclusions oriented parallel to two directions in the diopside crystal. RJL3519

Slabs of hedenbergite skarn from Dal'negorsk, Russia, are occasionally polished to show attractive, intricate growths of needle-like phases in contrasting colors.

The transparent varieties of spodumene, *hiddenite* (green) and *kunzite* (pink), have been known since the late 1800s. At that time, gem-quality spodumene was known only from locales in the United States. Modern sources of these gemstones now include pegmatites in Afghanistan, Madagascar, Brazil, and California.

Figure 10. Slab of hedenbergite skarn from Dal'negorsk showing banded growth of dark green hedenbergite and off-white wollastonite. RJL3531.

An interesting historical account of the discovery of *hiddenite* is given by Bauer (1904): "In North Carolina the hiddenite variety is found at Stony Point, the name of which place has been altered to Hiddenite, associated with emerald, beryl, quartz, garnet, rutile, and other minerals, in drusy cavities in gneissose granite. The first specimens of hiddenite were discovered in 1879 by Mr. W. E. Hidden; they had been weathered out of the mother-rock and were lying loosely on the ground. They were transparent and greenish-yellow in color and were at first thought to be diopside, since at that time spodumene had never been found in fine, transparent crystals. ... Some crystals are remarkable for the possession of a peculiar corroded surface." Based on its appearance and chemical composition, *hiddenite* was occasionally called "lithium emerald" (Kunz 1892).

The variety *kunzite* is named for George F. Kunz, who first described the gem in 1902. He also reported interesting experiments on its properties of thermoluminescence as well as fluorescence under ultraviolet light, X-rays, and exposure to various radioactive compounds (Baskerville and Kunz 1904).

In terms of sheer quantity and historical importance, jadeite is the most important gem mineral in the group and certainly most familiar to the layman. Many books have been written about jade from diverse perspectives including its use by modern jewelry and lapidary artists, its role in ancient cultures, and its significance to the antique and collectible art markets. Jadeite was prized by the Aztec, Olmec, and Mayan cultures in Central America for several thousand years, but after the Spanish conquest in the sixteenth century the sources of jadeite were lost for almost five hundred years. Small quantities were found in Guatemala beginning around 1950, and more concerted prospecting efforts uncovered blocks of jadeite in a serpentinite body in the Motagua river valley about 80 km northeast of Guatemala City (Hargett 1990). In the mid-1990s, large finds of jadeite were reported. The material was found as large boulders containing gem-grade jadeite, along with omphacite and other minerals. In addition to high-grade gem material, the large quantity of some grades makes it feasible to consider quarrying them as dimension stone (Cleary and Rohtert 2002).

An interesting jade-like material called *mawsitsit* is found in Myanmar (Burma) near the mines where imperial jadeite has traditionally been recovered. The major mineral phase in mawsitsit is kosmochlor, an extremely rare Na-Cr pyroxene first described from the Toluca, Mexico, meteorite. Mawsitsit is a colorful mixture of black and emerald green and takes a high polish, making it popular for cabochons and carvings.

Figure 11. Three small gemmy spodumene crystals. Left: *hiddenite* from Arassuai, Minas Gerais, Brazil; center: *kunzite* from Galileia, Minas Gerais, Brazil; right: *hiddenite* from Hiddenite, North Carolina.

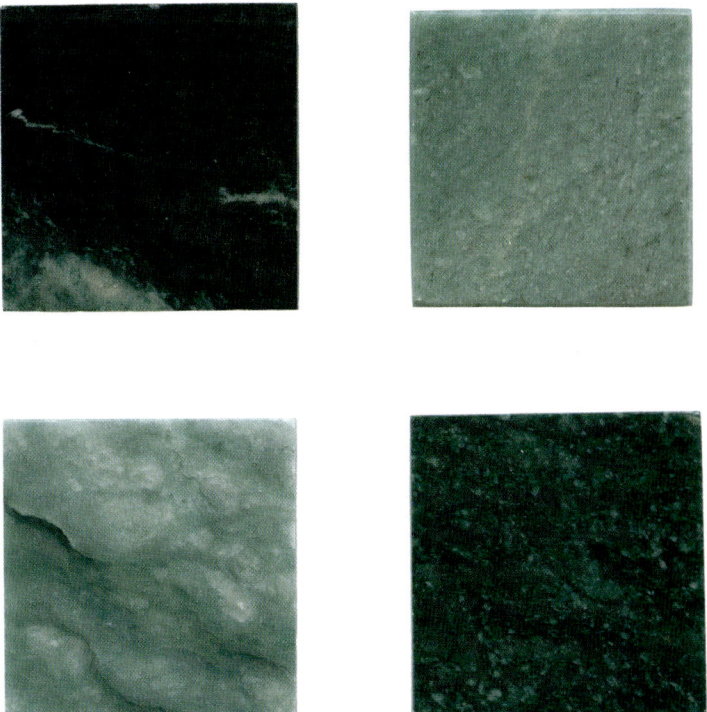

Figure 12. Sample tiles, each 5 × 5 cm, of several grades of material from the Guatemalan jade deposit. Clockwise from upper left: black; light green; aguacate, and jaguar. *Specimens courtesy of Ventana Mining Company, Reno, Nevada.*

# Taxonomy of the Pyroxene Group

## General Formula and Subdivisions of the Group

Pyroxenes are orthorhombic or monoclinic silicates with the general formula $M2M1T_2O_6$ where $M2$ is Ca, $Fe^{2+}$, Li, Mg, $Mn^{2+}$, Na; $M1$ is Al, $Cr^{3+}$, $Fe^{2+}$, $Fe^{3+}$, Mg, $Mn^{2+}$, Zn, $Sc^{3+}$, Ti, Zr, $V^{3+}$; and $T$ is Si, Al, $Fe^{3+}$. Here, $M2$ refers to cations in a generally distorted octahedral coordination, $M1$ are cations in a regular octahedral coordination, and $T$ are cations in a tetrahedral coordination (Morimoto et al. 1989). The group comprises roughly twenty minerals, arranged into six subdivisions:

## Mg-Fe pyroxenes

The **Mg-Fe pyroxenes** include the orthorhombic minerals enstatite and ferrosilite and the monoclinic minerals clinoenstatite, clinoferrosilite, and pigeonite. A classification system for orthopyroxenes with compositions intermediate between enstatite (Es) and ferrosilte (Fs) was first proposed by Poldervaart (1947) in which progressively more iron-rich members were designated *bronzite*, *hypersthene*, *ferrohypersthene*, and *eulite*. This scheme was widely adopted for many years, but under currently accepted nomenclature, only the end members are given species status and in general the "50% rule" is used to assign one of the end-member names to intermediate compositions in a solid solution series. However, collectors will often encounter these terms, particularly in connection with certain classes of meteorites and some interesting lapidary materials.

## Mn-Mg pyroxenes

The **Mn-Mg pyroxenes** include the orthorhombic species donpeacorite, $(Mn^{2+},Mg)MgSi_2O_6$, and its monoclinic dimorph, kanoite.

## Ca pyroxenes

The **Ca pyroxenes** are monoclinic silicates with the general formula $CaM1Si_2O_6$ where $M1$ = Mg, $Fe^{2+}$, $Mn^{2+}$, Zn, or $Fe^{3+}$. Recognized species are diopside, hedenbergite, johannsenite, augite, petedunnite, and esseneite.

## Ca-Na pyroxenes

The **Ca-Na pyroxenes** are represented by omphacite, $(Ca,Na)(Mg,Fe^{2+},Fe^{3+},Al)(Si_2O_6)$, and aegirine-augite, $(Ca,Na)(Mg,Fe^{2+},Fe^{3+})Si_2O_6$.

## Na pyroxenes

The **Na pyroxenes** are monoclinic minerals with the general formula $NaM1(Si_2O_6)$, where $M1$ is Al, $Fe^{3+}$, $Cr^{3+}$, $Mn^{3+}$, or $Sc^{3+}$. Recognized species include jadeite, aegirine, kosmochlor, namansilite, and jervisite.

## Li pyroxenes

The **Li pyroxenes** are represented by spodumene, $LiAlSi_2O_6$.

Accepted species and their ideal formulas are given in Table 1. Obsolete and varietal names, Table 2, were compiled largely from Morimoto et al. (1989), along with other sources.

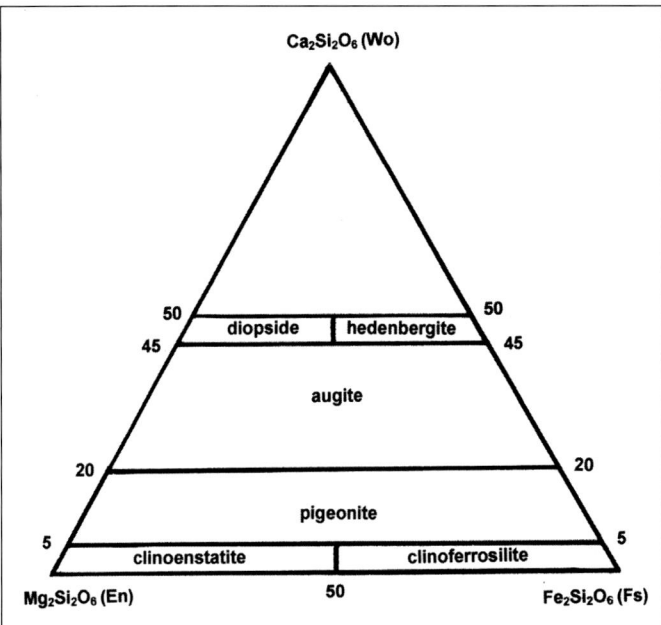

Figure 13. Composition range of the Ca-Mg-Fe clinopyroxenes, with accepted names (Morimoto et al. 1989).

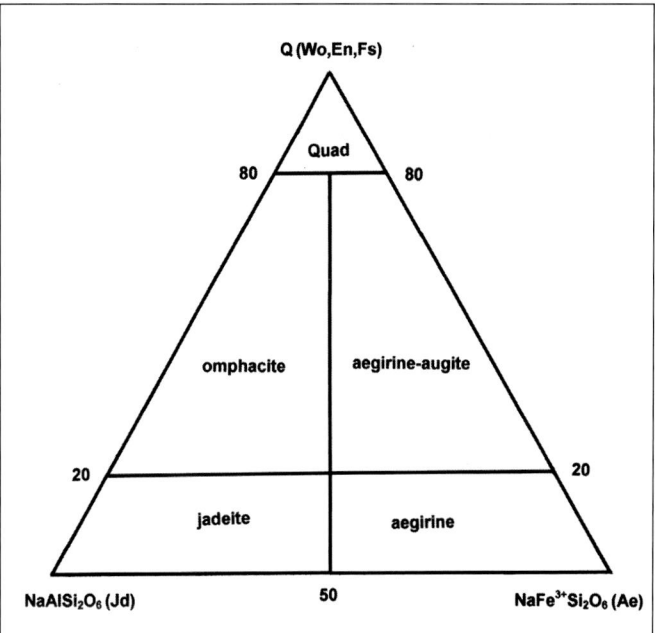

Figure 15. Composition ranges of the Ca-Mg-Fe and Na pyroxenes, with accepted names. Quad represents the Ca-Mg-Fe pyroxene area (see Figure 13) (Morimoto et al. 1989).

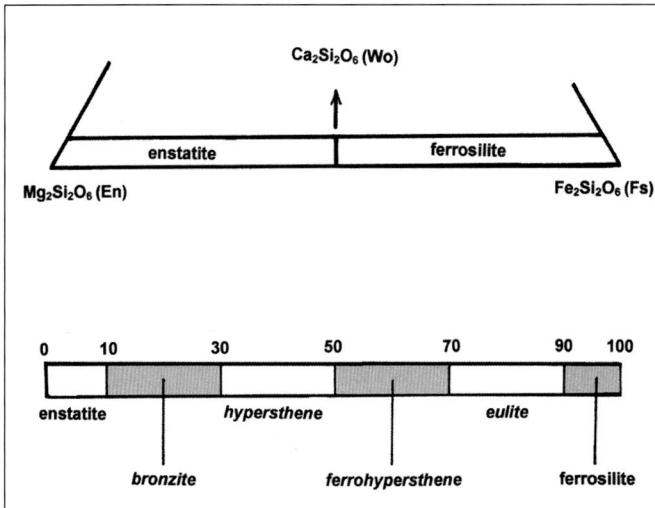

Figure 14. Top: Composition ranges of the orthopyroxenes, with accepted names (Morimoto et al. 1989). Bottom: Composition ranges corresponding to the varietal names formerly applied to intermediate compositions in the enstatite-ferrosilite series.

# Table 1. Accepted species and their formulas

| Species | Formula | Symmetry |
|---|---|---|
| Aegirine | $NaFe^{3+}Si_2O_6$ | monoclinic |
| Aegirine-augite | $(Ca,Na)(R^{2+},Fe^{3+})Si_2O_6$ | monoclinic |
| Augite | $(Ca,Mg,Fe)_2Si_2O_6$ | monoclinic |
| Clinoenstatite | $Mg_2Si_2O_6$ | monoclinic |
| Clinoferrosilite | $Fe_2Si_2O_6$ | monoclinic |
| Diopside | $CaMgSi_2O_6$ | monoclinic |
| Donpeacorite | $(Mn,Mg)MgSi_2O_6$ | monoclinic |
| Enstatite | $Mg_2Si_2O_6$ | orthorhombic |
| Esseneite | $CaFe^{3+}AlSiO_6$ | monoclinic |
| Ferrosilite | $Fe^{2+}_2Si_2O_6$ | orthorhombic |
| Hedenbergite | $CaFe^{2+}Si_2O_6$ | monoclinic |
| Jadeite | $NaAlSi_2O_6$ | monoclinic |
| Jervisite | $NaSc^{3+}Si_2O_6$ | monoclinic |
| Johannsenite | $CaMnSi_2O_6$ | monoclinic |
| Kanoite | $MnMgSi_2O_6$ | monoclinic |
| Kosmochlor | $NaCr^{3+}Si_2O_6$ | monoclinic |
| Namansilite | $NaMn^{3+}Si_2O_6$ | monoclinic |
| Natalyite | $Na(V^{3+},Cr)Si_2O_6$ | monoclinic |
| Omphacite | $(Ca,Na)(R^{2+},Al)Si_2O_6$ | monoclinic |
| Petedunnite | $CaZnSi_2O_6$ | monoclinic |
| Pigeonite | $(Mg,Fe,Ca)_2Si_2O_6$ | monoclinic |
| Spodumene | $LiAlSi_2O_6$ | monoclinic |

# Table 2. Obsolete and varietal names

| Obsolete Name | Correct Name |
|---|---|
| acmite | **aegirine** |
| aegirine-hedenbergite | **augite** |
| agalite | probably **enstatite** partially altered to talc |
| aglaite | altered **spodumene** |
| alalite | **diopside** |
| alkali-augite | **aegirine-augite** |
| amblystegite | **enstatite** |
| anthochroite | **augite** |
| asteroite | Fe-rich **augite** |
| baikalite | **diopside** |
| bastite | **enstatite** altered to serpentine or talc |
| blanfordite | Mn-rich **aegirine-augite** |
| bronzite | **enstatite** compositions from $Es_{90}Fs_{10}$ to $Es_{70}Fs_{30}$ |
| calc-clinobronzite | **pigeonite** |
| calc-clinoenstatite | **pigeonite** |
| calc-clinohypersthene | **pigeonite** |
| canaanite | **diopside** |
| chladnite | **enstatite** |
| chloromelanite | **omphacite** or **aegirine-augite** |
| chrome-acmite | Cr-rich **aegirine** |
| clinoeulite | Mg-rich **clinoferrosilite** |
| clinohypersthene | **clinoenstatite** or **clinoferrosilite** |
| coccolite | granular Fe-rich **diopside** |
| cosmochlore | **kosmochlor** |
| cymatolite | altered **spodumene** |
| diaclasite | altered **enstatite** |
| diallage | **diopside** that is altered or shows good (100) parting |
| diopside-jadeite | **omphacite** |
| endiopside | Mg-rich **augite** |
| eulite | **ferrosilite** compositions from $Es_{30}Fs_{70}$ to $Es_{10}Fs_{90}$ |
| eulysite | **ferrosilite** |
| fassaite | Fe- and Al-rich **diopside** or **augite** |
| fedorovite | **diopside** |
| ferroaugite | **augite** |
| ferrohedenbergite | **augite** |
| ferrohypersthene | **ferrosilite** compositions from $Es_{50}Fs_{50}$ to $Es_{30}Fs_{70}$ |
| ferro-johannsenite | Fe-rich **johannsenite** |
| ferropigeonite | Fe-rich **pigeonite** |
| ferrosalite | **hedenbergite** |
| ficinite | **enstatite** |
| funkite | **hedenbergite** |
| germarite | altered **enstatite** |
| hiddenite | green (gem) **spodumene** |
| hudsonite | **hedenbergite** |

| Obsolete Name | Correct Name |
|---|---|
| hypersthene | **enstatite** compositions from $Es_{70}Fs_{30}$ to $Es_{50}Fs_{50}$ |
| jadeite-aegirine | **jadeite** or **aegirine** |
| jeffersonite | Zn- and Mn-rich **diopside** or **augite** |
| killinite | altered **spodumene** |
| korea-augite | **augite** |
| kunzite | pink (gem) **spodumene** |
| lavroffite | **diopside** |
| leucaugite | **diopside** |
| lime-bronzite | "inverted" **pigeonite** or **enstatite** + **augite** |
| loganite | **diopside** + actinolite + talc |
| lotalite | **hedenbergite** |
| malacolite | **diopside** with good (100) parting |
| mansjoite | **augite** or **diopside** or **hedenbergite** |
| mayaite | **omphacite** |
| mellcrite | **orthopyroxene** |
| mondradite | altered **pyroxene** |
| mussite | **diopside** |
| orthobronzite | **enstatite** |
| orthoenstatite | **enstatite** |
| orthoeulite | **ferrosilite** |
| orthoferrosilite | **ferrosilite** |
| orthohypersthene | **enstatite** or **ferrosilite** |
| paulite | **enstatite** |
| peckhamite | **enstatite** |
| phastine | altered **enstatite** |
| picrophyll | altered **pyroxene** (?) |
| pigeonite-augite | probably subcalcic **augite** |
| pitkarantite | **pyroxene** (?) |
| protheite | **augite** |
| protobastite | **enstatite** |
| pyrallolite | altered **pyroxene** (?) |
| pyrgom | **pyroxene** |
| sahlite | **diopside** |
| salite | **diopside** |
| schefferite | Mn-rich **diopside** |
| schillerspar | **enstatite** altered to talc, serpentine, or anthophyllite |
| shepardite | enstatite |
| soda-spodumene | Na-rich **spodumene** |
| strakonitzite | altered **pyroxene** (?) |
| szaboite | partially altered **enstatite** |
| titanaugite | Ti-rich **augite** |
| titandiopside | Ti-rich **diopside** |
| titanpigeonite | Ti-rich **pigeonite** |
| trachyaugite | **augite** |
| traversellite | **diopside** |
| triphane | **spodumene** |
| tuxtlite | **omphacite** |
| uralite | pseudomorph of amphibole after diopside |
| urbanite | Fe-rich **augite** or **aegirine-augite** |
| ureyite | **kosmochlor** |
| vanadinaugite | V-rich **augite** |
| vanadinbronzite | V-rich **enstatite** |
| vargasite | altered **pyroxene** (?) |
| victorite | **enstatite** |
| violaite | **augite** |
| violan | Mn-rich **diopside** or **augite** |

## Crystal Structure and Morphology

The first pyroxene structure to be determined by X-ray diffraction methods was that of diopside. Warren and Bragg (1928) showed that the essential feature of the structure involves single chains of $SiO_4$ tetrahedra, leading to an ($SiO_3$) repeat unit. By the 1930s, the structures of most of the pyroxenes had been worked out. The structures of the closely related "pyroxenoids" pectolite and wollastonite were not solved for another thirty years. The pyroxenoids are also single-chain silicates and have some structural similarities to the pyroxenes as well as several key differences (Prewitt and Peacor 1964).

A polyhedral model of the structure of diopside, viewed along the x-axis, is shown in Figure 16. It can be seen that each silicate tetrahedron shares one oxygen ion with each of its neighbors to create the continuous chain. The "backbone" of the chain, an imaginary line drawn through all the shared oxygens, isn't perfectly straight. By comparing structure refinements of six natural pyroxenes and two synthetic compounds having the pyroxene structure, it was found that the linearity of the silicate chain varies somewhat, based on the relative sizes of the other metal ions that are present, with jadeite and aegirine having the straightest chains. The greatest departures from linearity are seen in the Ca pyroxenes, diopside and johannsenite (Clark, Appleman, and Papike 1969).

The structures of clinopyroxene and orthopyroxene viewed along the y-axis are shown in Figures 17 and 18, illustrating the oblique angle between the x- and z-axes in the clinopyroxenes versus the orthogonal relationship in the orthopyroxenes.

The relationship between the composition of a pyroxene crystal and its cell dimensions was analyzed by Smith et al. (1969) who concluded that cell dimensions as measured by X-ray diffraction are not a reliable guide for predicting Mg, Ca, and Al content. Because of the importance of pyroxenes to many problems in petrology, much research has been done to examine their structures at high and low temperatures and pressures (see, e.g., Cameron et al. 1973; Finger and Osashi 1976; Prencipe et al. 2000; Levien and Prewitt 1981).

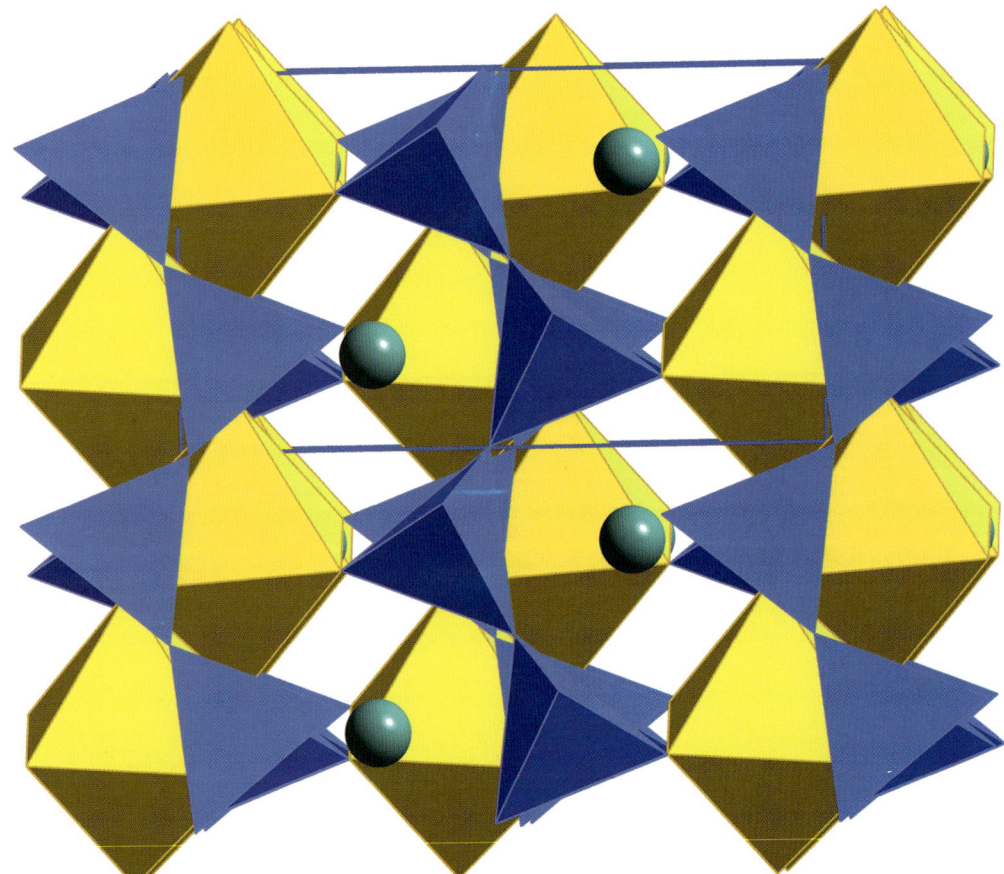

Figure 16. The crystal structure of diopside viewed along the x-axis, showing the chains of $SiO_4$ tetrahedra (dark blue) running parallel to the z-axis. Mg atoms are at the centers of the yellow octahedra, and green balls represent Ca. The unit cell is indicated by blue lines.

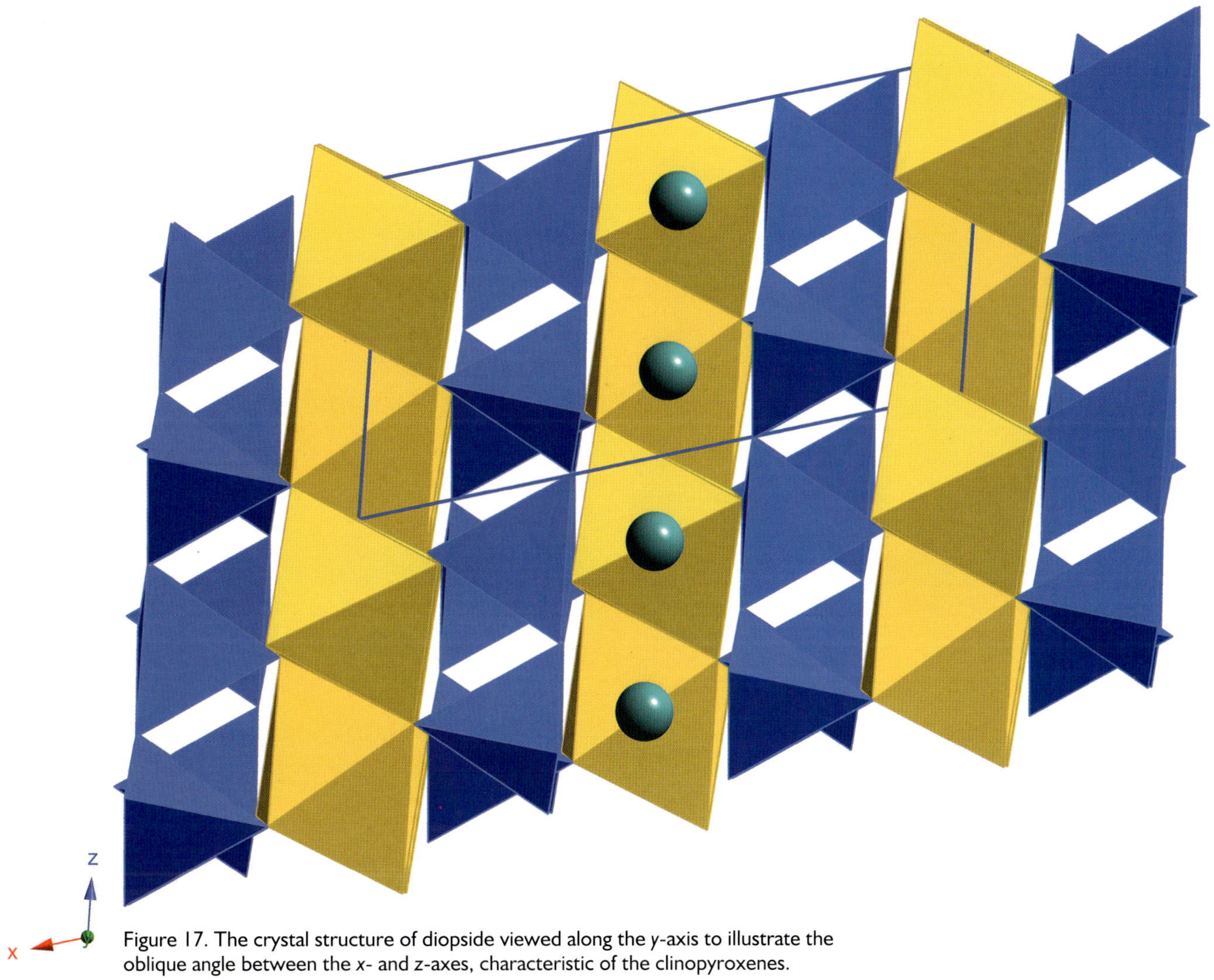

### Taxonomy of the Pyroxene Group: Crystal Structure and Morphology

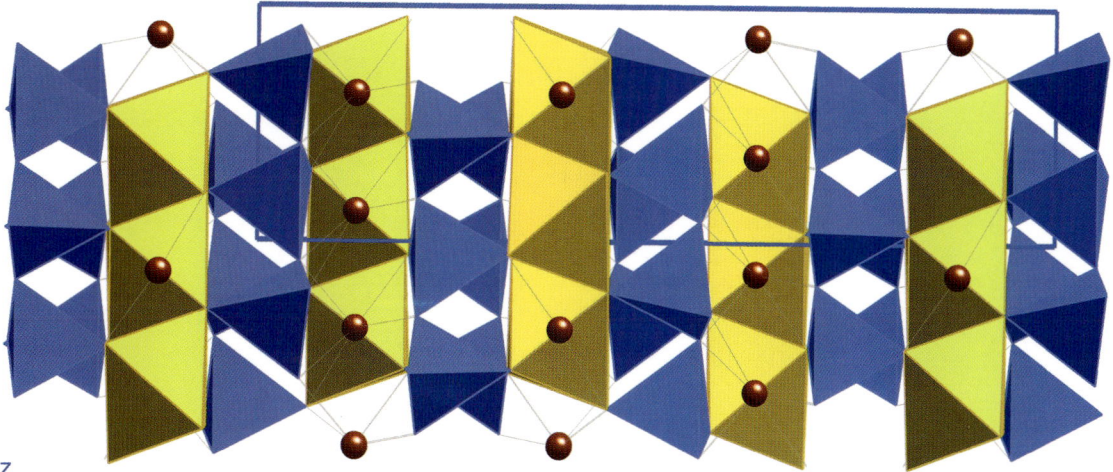

Figure 17. The crystal structure of diopside viewed along the y-axis to illustrate the oblique angle between the x- and z-axes, characteristic of the clinopyroxenes.

Figure 18. The general structure of an orthopyroxene, viewed along the y-axis, showing the orthogonal relationship between the x- and z-axes (compare to preceding Figure). Brown balls represent $Fe^{2+}$ and yellow octahedra represent Mg.

## Taxonomy of the Pyroxene Group: Crystal Structure and Morphology

Pyroxenes can be found in a wide variety of interesting habits, ranging from equant to tabular crystals, flattened to fanlike aggregates, radiating acicular groups, and elongated prismatic crystals with acute terminations. Goldschmidt (1922) presented 67 drawings of orthopyroxenes and over 600 drawings of clinopyroxenes. (These numbers don't include the illustrations of several species that Goldschmidt grouped with the pyroxenes but that are now considered pyroxenoids.)

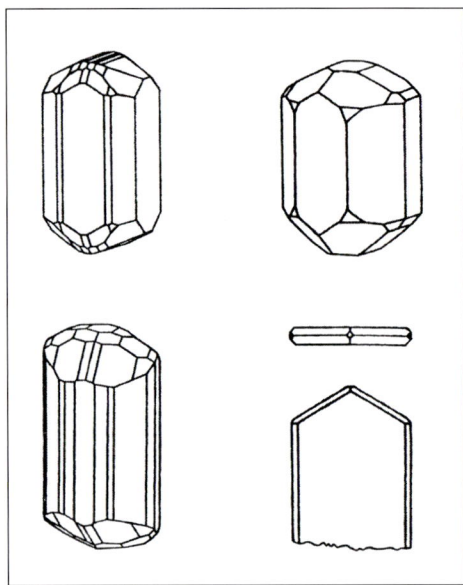

Figure 19. Morphological drawings of some orthopyroxenes, modified from Goldschmidt (1922). Top: equant enstatite crystals from Mt. Dore, France (left) and Oedegarden, Norway (right). Bottom: tabular enstatite var. *bronzite* from a German meteorite (left) and enstatite from Siebenbürgen (right).

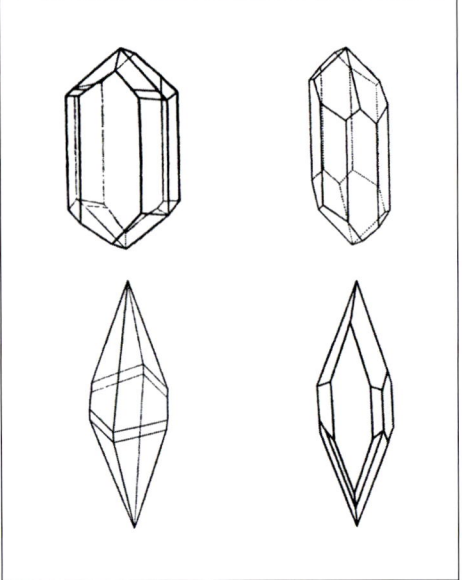

Figure 20. Morphological drawings of some clinopyroxenes, modified from Goldschmidt (1922). Top: equant diopside var. *jeffersonite* from Sparta, New Jersey (left) and augite from Pargas, Finland (right). Bottom: Diopside var. *fassaite* from Fassa, Tyrol (left) and aegirine from near Eker, Norway (right).

Figure 21. Diopside crystal from Otter Lake, Quebec, Canada, showing stout euhedral habit. The crystal is about 3 cm tall. *RJL3319*

## Taxonomy of the Pyroxene Group: Crystal Structure and Morphology

Figure 22. Hedenbergite from the Central Bor mine, Dal'negorsk, Russia, forming sprays of subparallel, flattened crystals. Specimen is 3 cm tall overall. *RJL2907*

Figure 23. Another hedenbergite from Dal'negorsk; in this case the hedenbergite crystals are extremely thin needles associated with small colorless quartz crystals. Specimen is a little over 4 cm tall. *RJL2785*

# Crystal Chemistry

Important topics in pyroxene crystal chemistry include the extent of solid solutions between the various species, the kinds of coupled substitutions that can take place (particularly the extent that Al can substitute for Si on the tetrahedral sites), and the formation of exsolution structures. Cameron and Papike (1981) reviewed the chemistry and structures of 175 naturally occurring pyroxenes from a variety of rock types and presented a number of observations:

1. A complete solid solution exists between diopside and hedenbergite and extensive solid solution exists between enstatite and ferrosilite under crustal conditions.
2. The total Al ranges between 0 and 1.0, which was lower than expected, and the maximum amount of Al substitution on the tetrahedral sites was 55% of **T** site occupancy.
3. In terrestrial Fe-Mg pyroxenes and augites, the most important substitutional couples are $Fe^{3+}$ on octahedral sites coupled with $Al^{3+}$ on tetrahedral sites, and $Al^{3+}$ on octahedral sites coupled with $Al^{3+}$ on tetrahedral sites. In planetary basalts, the coupled substitution of octahedral $Ti^{4+}$ with tetrahedral $Al^{3+}$ is also important.
4. Transmission electron microscope (TEM) studies of the textures of exsolution lamellae have helped to clarify the mechanisms (nucleation and growth versus spinodal decomposition) by which exsolution occurs.
5. Some features in pyroxene crystals that are potentially useful as geothermometers include the Fe-Mg intracrystalline distribution, the orientation of exsolution lamellae relative to the (001) and (100) planes of the host crystal, and differential changes that occur in the cell parameters of the host and lamellar phases during cooling.

The incorporation of hydroxyl ($OH^-$) in pyroxenes is of interest because it can provide insights into the role of water in the Earth's upper mantle. Some of the more common hydrous silicates such as phlogopite and members of the amphibole group are not expected to exist at great depths, and the traces of water found in other minerals cannot account for the amount of water believed to be present in the mantle. Although pyroxenes are normally regarded as anhydrous minerals, infrared spectroscopy has shown that a sample of omphacite from an eclogite nodule from the Roberts Victor kimberlite pipe, South Africa, contains up to 1000 ppm $OH^-$, suggesting that such pyroxenes might be a significant reservoir for water in the upper mantle (Skogby, Bell, and Rossman 1990; Smyth, Bell, and Rossman 1991).

# Formation and Geochemistry

## Pyroxenes in Igneous Rocks

Pyroxenes are common constituents of basalts and gabbros. Enstatite is an important component of many ultrabasic and ultramafic rocks. Clinoferrosilite is found in the lithophysae of obsidians at several locales including Lake Naivasha, Kenya, and Coso Hot Springs, California. Pigeonite is a characteristic component of andesites and dacites. The clinopyroxenes that crystallize early in basic magmas tend to be augite, but diopside can also be formed in this way. Sharp augite crystals are found in volcanic rocks, for example on Mt. Teide, Canary Islands.

For collectors, many excellent specimens are found in pegmatites. For example, fine aegirine crystals are associated with feldspar and zircon in alkali pegmatites at Mt. Malosa, Malawi. Spodumene (especially var. *kunzite*) is a characteristic mineral of lithium pegmatites; it is of interest because it is a major ore of lithium and it plays an important role in the crystallization of complex Li-rich pegmatites. Exceptional gemmy spodumenes of various colors in huge sizes (up to 60 cm long!) have been found in the pegmatites of Laghman, Afghanistan (Bariand and Poullen 1978). Similar pegmatite occurrences include those of San Diego County, California, and numerous places in Brazil. Giant spodumene crystals occur in pegmatites in the Black Hills, Wyoming, and in Taos County, New Mexico.

Figure 24. Pink crystal of spodumene var. *kunzite* from Mawi, Laghman, Afghanistan, illustrates a typical pegmatite association of spodumene with off-white microcline. The crystal is about 10 cm tall. *RJL1126*

Aegirine and aegirine-augite are characteristic components of some alkaline igneous rocks, particularly syenites. Some noteworthy examples are found at Mont Saint-Hilaire, Quebec; in alkaline complexes on the Kola Peninsula, Russia; the Narssarssuk pegmatite, Greenland; and at Magnet Cove, Arkansas.

Figure 25. Aegirine crystals about 3 cm tall on feldspar from Mount Malosa, Malawi. *RJL1545*

## Pyroxenes in Metamorphic Rocks

At the Yates mine, Otter Lake, Quebec, Canada, highly mineralized skarns of the Grenville marble host fine, large crystals of diopside and other pyroxenes (mostly identified as augite) (Leavitt 1981). Hedenbergite is found in many habits in the skarns at Dal'negorsk, Russia, often forming spectacular specimens with interesting associated species (Lisitsyn and Malinko 1994; Moroshkin and Frishman 2001; Grant and Wilson 2001). Skarn deposits at Serifos, Greece, are notable for hedenbergite in several habits and associations including columnar green crystals in subparallel arrangements, microscopic green inclusions in quartz, and colorful associations with andradite garnet (Gauthier and Albandakis 1991).

Figure 26. Sharp, greenish black pyroxene crystal about 3 cm tall, with orange calcite and minor purple fluorite from Otter Lake, Quebec, Canada. Most of the pyroxenes at this locale are diopside, with lesser amounts of augite and possibly others reported. This specimen was identified by the seller as ferrosilite, but in the absence of a reliable analysis any identification must be considered tentative, and in this case the crystal is more likely diopside. *RJL2919*

Figure 27. A cabinet-sized specimen illustrating a classic association from the skarns at Dal'negorsk, Russia: lustrous black ilvaite crystals to 25 mm long, with milky bipyramidal quartz on a massive matrix of bladed brownish green hedenbergite, from the Sovietskiy #1 mine. *RJL3196*

**Formation and Geochemistry: Pyroxenes in Metamorphic Rocks   27**

Figure 28. Another typical skarn assemblage: orange-brown andradite garnet and green hedenbergite from Dal'negorsk. Specimen is about 5 cm wide. *RJL3397*

Figure 29. Photo of andradite crystals to about 1 cm on hedenbergite from Serifos, Greece. *RJL2128*

Several important diopside occurrences lie within a belt of serpentinized ultramafic rocks that runs northeast from the asbestos deposit at Eden Mills, Vermont, into Canada, and includes the Orford nickel mine, the Jeffrey mine, and asbestos mines at Black Lake and Thetford Mines, all in Quebec (Tarassof and Gault 1994; Grubb 1965).

Eclogites are metamorphic rocks that represent compositions similar to those of a typical basic igneous rock that has been subjected to metamorphism at high temperatures and pressures. Omphacite is the characteristic pyroxene in eclogite and it is typically associated with equiaxed garnet crystals (typically pyrope). The resulting colorful combination is occasionally used as a lapidary material or ornamental stone.

Aegirine, in dark green to black radiating masses, is a minor accessory mineral in charoite, an attractive lavender ornamental mineral from the Aldan Shield, Russia; the deposit is interpreted as a K-feldspar metasomatite formed

Figure 30. Tiny dark green Cr-rich grossular crystals on tan laths of diopside from the Orford nickel mine, Quebec, Canada. Overall length is about 6 cm. Collectors should be aware that older labels occasionally misidentify the locale as "Oxford, Quebec" and misidentify the garnet as uvarovite. *RJL156*

at the contact of aegirine- and nepheline-syenites with limestones (Rogova et al. 1978)

Pyrometamorphism is a special type of thermal metamorphism caused by local combustion of coal, wood, or other carbonaceous material. The phenomenon can arise through completely natural processes or as a result of human activities. During their travels, Lewis and Clark saw naturally burning coal seams and realized the connection to some distinctive features of the landscape. Combustion can arise when a coal seam is exposed to air through erosion or a falling water table; ignition can be the result of lightning, brush fire, or simple oxidation (spontaneous combustion). As the coal burns, the overburden slumps or collapses, creating more avenues for air to reach the coal. It is estimated that over the last two million years, the amount of coal consumed by natural fires in the Powder River Basin of Wyoming and Montana is *one to two orders of magnitude greater* than the total amount of coal mined there in the past century (Heffern and Coates 2004). Heat from the burning coal seam sinters the adjoining sedimentary rock into a hard red or multicolored rock called a "clinker." Where the temperatures are highest (~1300°C), the rock can actually be melted, and the solidified melt is called "paralava."

Figure 31. A polished section of lavender charoite containing radiating masses of greenish black aegirine, from the Murun massif, Russia. *RJL2526*

Outcrops of these formations cover roughly 37,000 km² in eastern Wyoming alone (Clark and Peacor 1992; Cosca and Peacor 1987). Detailed studies of the minerals in a silica-undersaturated paralava from the Powder River Basin, near Gillette, Wyoming, led to the description of the new clinopyroxene species esseneite. The mineral is associated with melilite, anorthite, magnetite-hercynite solid solution, and some glass (Cosca and Peacor 1987).

In the Chelyabinsk coal basin, southern Urals, Russia, spontaneous combustion has taken place in heaps of waste rock from mining activities. There are about fifty such heaps in the Chelyabinsk district, and they are typically 40–70 m tall and contain as much as 1,000,000 m³ of material. The waste heaps began as a mixture of mudstones, siltstones, siderite concretions, sandstone, and pieces of ancient wood, along with some coal. The carbonaceous material began smoldering and combustion lasted for ten to fifteen years, with localized areas sometimes passing through an actual flame combustion stage. The skarn-like material that resulted was estimated to have formed at temperatures of at least 1000°C and cooled to ~300–500°C over days to weeks. Small (~1 mm) prismatic pyroxene crystals were noted, associated with anorthite, hematite, melilite, ferromagnesiohastingsite, and anhydrite (Kabalov et al. 1997; Sokol, Volkova, and Lepezin 1998).

Paralava can also be formed where wood is burned under conditions that subject nearby rocks to intense heat. Heat from burning gorse stacked on a volcogenic graywacke near Colac Bay, New Zealand, melted the surface of the graywacke at ~1330°C; the resulting glassy paralava contained Al-rich diopside (Coombs et al. 2008). Similar occurrences in Europe have been attributed to wood burning to make charcoal.

Figure 32. A typical sample of paralava from Kopiesk, Southern Urals, Russia. Specimen is about 3 cm wide.

## Pyroxenes in Extraterrestrial Rocks

Pyroxenes from extraterrestrial sources include those found in dust particles collected from space, chondritic meteorites, achondritic meteorites, iron meteorites, and rock samples collected during lunar exploration. The chemistry and texture of the pyroxene minerals in these specimens has contributed greatly to our understanding of their origins and history. Mason (1968) notes: "Pyroxenes are almost ubiquitous in meteorites. Among the stones and stony-irons, only the pallasites and a few carbonaceous chondrites do not contain pyroxenes, and even the iron meteorites sometimes have pyroxene-bearing silicate inclusions."

Dust particles are the first material to solidify out of the primordial gases created in stars. Particles collected during the Stardust spacecraft mission to the comet Wild 2 (*aka* comet 81P) in 2004 were analyzed by electron microscopy, X-ray diffraction, and other advanced techniques. Many of the particles contained pyroxene, some of which was diopside, but enstatite was more commonly found (A'Hearn 2006; Brownlee et al. 2006; Keller et al. 2006; Zolensky et al. 2006).

Chondritic meteorites represent fragments of primitive material that has undergone accretion but not the differentiation or magmatism associated with larger parent bodies.

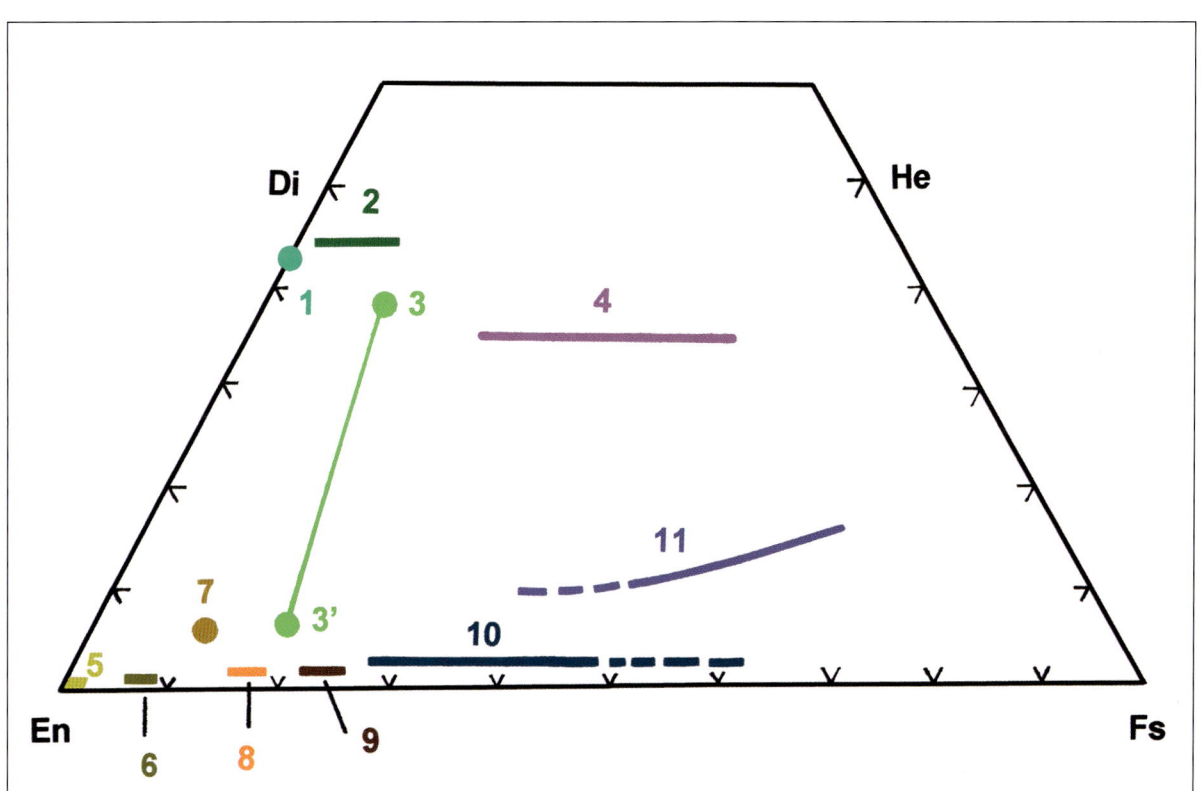

Figure 33. Composition ranges for pyroxene in several types of meteorites [after Mason (1968)]:
1 diopside in estatite achondrite
2 diopsides in chondrites and silicate inclusions in irons
3-3' coexisting augite and orthopyroxene in chondrite
4 augites in mesosiderites and achondrites
5 enstatite and clinoenstatite in enstatite chondrites and enstatite achondrites
6 enstatite in silicate inclusions in irons
7 clinoenstatite in ureilite
8 orthopyroxene in bronzite chondrites
9 orthopyroxene in hypersthene chondrite
10 orthopyroxene in achondrites and stony-iron mesosiderites
11 pigeonite in achondrites and mesosiderites

Orthopyroxene is a major constituent of many chondritic meteorites. Important groups of noncarbonaceous chondrites include: *enstatite chondrites* having pyroxene compositions near the enstatite end member (i.e., $Fs_{\approx 0}$); *olivine-bronzite chondrites* with pyroxene compositions $Fs_{15-18}$; *olivine-hypersthene chondrites* with pyroxene compositions $Fs_{20-23}$; and *amphoterites* (or amphoteric chondrites) with pyroxene compositions $Fs_{25-27}$, where Fs = ferrosilite (Binns 1970). Chondritic orthopyroxenes, compared to those in terrestrial rocks, tend to be poorer in $Fe^{3+}$ and Ca. Enstatite or clinoenstatite is the essential constituent in enstatite chondrites and, as noted, in many cases it is close to pure $Mg_2Si_2O_6$ as shown in Figure 33, in which data are plotted using the same compositional triangle as in Figures 13-14.

Achondritic meteorites include the eucrites, diogenites, and shergottites, along with several other rare types. *Eucrites* are superficially similar to terrestrial basalts; however, the pyroxene in eucrites tends to be pigeonite, whereas in terrestrial basalts augite is more commonplace. *Diogenites* are made up almost entirely of Ca-poor pyroxenes, associ-

Figure 34. A polished slice of the Beaver, Oklahoma, meteorite, an L5 olivine-hypersthene chondrite. The mineral phases include Ni-Fe, troilite, chromite, and pyroxene. Sample is about 2 X 4 cm. *RJL1175*

Figure 35. A cut slab of the Nuevo Mercurio, Mexico, meteorite, an H5 olivine-bronzite chondrite. Specimen is about 3 cm wide. *RJL1221*

Figure 36. A cut slice of the Correo, New Mexico, meteorite, an H4 olivine-bronzite chondrite. Analysis of the orthopyroxene component in this meteorite gave an average composition of $En_{82.3}Wo_{1.2}Fs_{16.5}$ (Rubin et al. 1981). Sample is about 3 cm wide. *RJL1006*

ated with small amounts of olivine or plagioclase feldspar. *Shergottites* are basaltic rocks consisting mainly of pyroxene and plagioclase, but in contrast to the eucrites, shergottites are more like terrestrial basalts: the pyroxene is Ca-rich augite and Ca-poor pigeonite, and the presence of magnetite indicates formation under more oxidizing conditions than the eucrites (McSween 1987).

Many eucrites are thought to be fragments of one differentiated asteroid (perhaps 4 Vesta), although other eucrites such as Ibitira, Brazil, and Northwest Africa 011 deviate significantly in their geochemical characteristics, suggesting that they were derived from different source bodies (Mittlefehldt 2005; Yamaguchi et al. 2002).

The Zagami, Nigeria, shergottite is a basalt that contains FeO-enriched pigeonite (composition range $\sim Fs_{27-80}$) as well as augite (composition range $\sim Fs_{19-35}$). Its chemistry and microstructure suggest that it is the product of fractional crystallization. Zagami is especially interesting because it is widely believed to have come from Mars (McCoy, Wadhwa, and Keil 1999; Wood and Ashwal 1981; McSween 1985).

Figure 37. A section of the Millbillillie, Australia, meteorite, an ordinary basaltic eucrite that fell in October 1960. Sample is about 3 cm wide. *RJL1234*

Figure 38. A well-preserved fragment of the Camel Donga, Australia, meteorite, a eucrite consisting predominantly of pyroxene ($\sim Fs_{49}Wo_{16}$) and plagioclase feldspar. The shiny black fusion crust with numerous contraction fractures is typical of specimens from this find (Cleverly, Jarosewich, and Mason 1986). Sample is about 4 cm long. *RJL1273*

Figure 39. Small fragments of the Zagami, Nigeria, eucrite. This meteorite is of particular interest because it is generally considered to be a fragment of Mars. Scale bar = 1 cm. *RJL2895*

Iron meteorites represent fragments of an object that was large enough to have undergone fairly complete differentiation into a silicate crust/mantle and a metallic core like that of the earth. The rare pyroxene kosmochlor was first described from an iron meteorite (the Toluca, Mexico octahedrite) in 1897 and then from the Coahuila and Hex River Mountains, Mexico hexahedrites (Frondel and Klein 1965), and has only been found in terrestrial rocks in the last twenty years or so. A detailed history of the Toluca meteorite and a summary of the scientific studies on it can be found in Buchwald (1975).

Pyroxenes are fairly abundant on the moon, as summarized in Table 3. Of the various rock types, basalts are perhaps most useful for comparing planetary histories because basaltic magma is derived from partial melting of the planetary interior and will therefore reflect the original accretionary material and the planet's evolutionary history. ("Planet" in this case means the earth, the moon, and the parent bodies of basaltic meteorites.) Papike (1980) notes: "Because pyroxenes are the most abundant ferromagnesian phase in most basalts and since their crystal structures accommodate all of the major elements that occur in basalts (except potassium) they, like their host basalts, are potentially powerful planetary probes."

The mineralogy of lunar basalts reflects some basic differences compared to the earth, particularly the low oxygen activity in the lunar environment, which reduces the valence states of several cations. Trivalent iron is virtually nonexistent on the moon, some Ti is reduced to $Ti^{3+}$, and some Cr is reduced to $Cr^{2+}$. Overall, the main differences between lunar mare basalts and terrestrial midocean ridge basalts can be attributed to the differing planetary characteristics: 1. mantle material that forms the source for midocean ridge basalts has a higher $Mg/(Mg+Fe^{2+})$ ratio and higher $Al_2O_3$ relative to the corresponding source material on the moon; 2. the moon is relatively depleted in alkali elements compared to the earth; and 3. lunar basalts developed under extremely low oxygen activity (Papike 1980). Pyroxenes identified in lunar igneous rocks from the Apollo 11 mission include augite, *ferroaugite*, pigeonite, and Ca-poor *ferroaugite* (Brown et al. 1970).

Figure 40. A polished slice of the Toluca, Mexico, iron meteorite, from which tiny crystals of the rare pyroxene kosmochlor were first isolated and described. *RJL3529*

# Table 3. Pyroxene abundances on the moon[a]

| Lunar material | Pyroxene, vol % |
|---|---|
| Mare basalts[b] | 40 to 65 |
| Anorthositic rocks[c] | 0 to 40 |
| Fragmental breccias[d] | 5 to 30 |
| Soils[e] | 5 to 20 |

[a] Data from Williams and Jadwick (1980)
[b] A few samples have less pyroxene
[c] The pyroxene is mostly Ca-poor in these rocks
[d] Value quoted for mineral grains 25 $\mu$m across
[e] Type and amount of pyroxene resembles the local rock (high-Ca in mare regions and low-Ca in highlands)

# The Minerals

## Aegirine

Aegirine is typically found in alkaline igneous rocks such as syenites, syenite pegmatites, and alkali granites. It is also occasionally found in crystalline schists, metamorphosed Fe-rich sediments, and hydrothermal and metasomatic deposits. It was first described by Berzelius in 1835 from Rundemeyr, Norway, and named for Aegir, the Scandanavian sea-god. Originally the term aegirine was applied to green-to-black crystals with blunt terminations, whereas *acmite* referred to brownish forms with sharply pointed terminations. In 1871, Tschermak demonstrated that aegirine and *acmite* are chemically identical. Under currently accepted nomenclature, *acmite* is considered an obsolete synonym for aegirine and should not be used.

In alkaline igneous rocks, aegirine is typically associated with amphiboles such as arfvedsonite and magnesioriebeckite, andradite, astrophyllite, aenigmatite, catapleiite, and låvenite. Aegirine is a late-crystallizing component, often mantling earlier-formed pyroxenes. In some rocks of the Lovozero massif, Russia, aegirine can make up as much as 20 to 35% of the rock (Deer, Howie, and Zussman 1978). At the Iron mine on the Kovdor massif, Kola Peninsula, Russia, it is found in alkaline syenite (ijolite) as black, elongated prismatic crystals and radiating aggregates in gray microcline (Ivanyuk and Yakovenchuk 1997). At Mont Saint-Hilaire, Quebec, Canada, "Aegirine is one of the most widespread minerals ... and occurs in pegmatites, sodalite syenite, silicate cavities, carbonate cavities, igneous breccia, hornfels, marble xenoliths, sodalite xenoliths, and as a rock-forming mineral in many of the igneous rocks. It is a common associate of almost every mineral which occurs there..." (Mandarino and Anderson 1989). Fine crystals are found in a pegmatite at Mount Malosa, Zomba district, Malawi, associated with microcline and zircon.

Aegirine has also been described from metamorphic rocks at many localities, often where sodium has been introduced through hydrothermal or metasomatic processes. In altered Fe-bearing shales and cherts in the Cuyuna district, Minnesota, the aegirine is interpreted to have formed from the reaction of hot, Na-rich waters on hematite cherts (Grout 1946). Aegirine was formed by sodium metasomatism of the Arunta feldspathic gneiss adjacent to metamorphosed carbonatite in the Strangeways Range, Northern Territory, Australia (Deer, Howie, and Zussman 1978).

Aegirine containing 0.41% $ZrO_2$ and 0.48% $Ce_2O_3$ is found at Quincy, Massachusetts. A Zr-rich variety was also reported from Kangerdlugssuaq, eastern Greenland. Aegirine containing 9.4% $TiO_2$ occurs in New South Wales, Australia. Based on a systematic study of "acmitic pyroxenes" from many locales, Washington and Merwin (1927) observed, "A striking and seemingly constant feature of acmite proper is the notable content of zirconia and the rare earths, the acmitic pyroxenes that contain much diopside or hedenbergite carrying but traces of these. This richness in zirconia and the rare earths is in harmony with the tendency of these oxides to be associated with soda in minerals and rocks."

**38** *The Minerals: Aegirine*

Figure 41. Aegirine with orthoclase from Mount Malosa, Malawi. Specimen is 8 cm tall. *RJL2363*

Figure 42. A group of large, terminated aegirine crystals, about 10 cm tall, with feldspar and minor zircon, from Mount Malosa, Malawi. *RJL3581*

Figure 43. A tan, doubly terminated zircon crystal about 2 cm long perched on black aegirine from Mount Malosa, Malawi. *RJL2362*

Figure 44. Small dark green aegirine crystals with siderite and natrolite from Mont Saint-Hilaire, Quebec, Canada.

## The Minerals: Aegirine 41

Figure 46. A dark aegirine crystal on a silvery rosette of polylithionite, from Mont Saint-Hilaire, Quebec, Canada. Sample is about 3 cm wide. *RJL3194*

Figure 45. Sharp aegirine crystal about 3 cm tall, from Mont Saint-Hilaire, Quebec, Canada. *RJL3586*

Figure 47. Radiating dark green aggregates of aegirine in charoite from the Murun massif, Russia. Sample is about 5 cm wide. *RJL2526*

42  The Minerals: Aegirine

Figure 48. Aegirine in nepheline syenite from Marathon County, Wisconsin. The acicular form of aegirine is sometimes referred to as *acmite*, a term that is now officially discouraged. *RJL203*

# Aegirine-augite

Aegirine-augite represents members of the series $Ca(Mg,Fe)Si_2O_6$-$NaFe^{3+}Si_2O_6$ that contain between $Fe^{3+}_{0.2}$ and $Fe^{3+}_{0.8}$ and lie in the region indicated in Figure 15. Although mineral collectors will rarely encounter aegirine-augite from dealers or see it commonly offered at mineral shows, it has been documented at over a hundred locales worldwide and occasionally forms impressively large crystals. For example, crystals up to one meter long and 35 X 20 cm in cross section have been described from a nepheline syenite adjacent to aegirinite, and in a coarse-grained ijolite-urtite pegmatite in the Kola Peninsula, Russia (Polkanov 1939).

Aegirine-augite is commonly formed in similar environments to those of aegirine, viz., alkali granites, quartz syenites, syenites, and nepheline syenites and their related pegmatites. Some documented occurrences, based on analyses tabulated in Deer, Howie, and Zussman (1978), include: in ijolite pegmatite in the Homa Bay area, Kenya; in foyaite at Kangerdlugssuaq, east Greenland; in pyroxene nepheline syenodiorite, Khamman district, India; in banded syenodiorite at Iron Hill, Colorado; in ijolite at Sealbrook Lake, Ontario, Canada; in syenite pegmatite in the Ilmen Mountains, Russia; and in alkali amphibole-garnet schist at Hodono, Bessi district, Japan.

Orange-yellow prismatic crystals up to about 1 cm long have recently been reported from basaltic scoria cones at two localities in Western Victoria, Australia (Mount Shadwell and Mount Anakie). The crystals typically grow directly on quartz, and are interpreted to have crystallized under high temperature conditions during post-eruptive cooling within the scoria mound (Birch et al. 2008). Detailed crystal structure analysis of material from Mount Anakie was reported by Mills and Groat (2008).

A Mn-rich variety, *blanfordite*, is found in regionally- and contact-metamorphosed manganiferous sediments at several localities in India, usually associated with the Mn-rich variety of magnesioriebeckite, *juddite*. Other associated minerals include Mn-rich arfvedsonite, Mn-rich chlorite, microcline, and braunite (Roy 1970; Roy 1971). It is interesting to note that *blanfordite* was first described in 1906 from the Kacharwahi manganese mines, Nagpur district, and similar material was reported from: Kachidhana, Chindwara district; Kajlidongri, Jhabua; and Jothvad, Narukot. It was later reported from Gangpur, from Ponia and Tirodi, Balaghat district, and from Chikla, Bandara district. Some occurrences represent regionally metamorphosed manganese silicate rocks (gondites), whereas others represent contact metamorphic rocks (kodurites) where manganese silicates have been penetrated by pegmatites and veins. Kilpady (1960) considered it to be a distinct mineral species, based on detailed analyses. Roy (1971) considered most *blanfordite* compositions to lie within the aegirine-augite field; however, *blanfordites* from the Precambrian Gangpur group of regionally metamorphosed Mn-rich sediments were identified as Mn-rich aegirine (Nayak, Mohapatra, and Sahoo 1997).

Figure 49. Reddish-purple masses and crude tabular crystals to about 2 cm long, of the Mn-rich variety of aegirine-augite, referred to as *blanfordite*, in metamorphosed manganiferous sediment, from the Tirodi mines, India. RJL3653

## Augite

Augite can be found in many types of igneous rocks, and is especially common in gabbros, dolerites, and basalts; it is also found less commonly in ultrabasic and intermediate rocks. Augite is occasionally found in regionally metamorphosed rocks, but these occurrences are quite rare compared to those in igneous rocks. Al-rich augite is sometimes found in metamorphosed limestones, associated with spinel; it can also occur at contacts between igneous and carbonate rocks, associated with garnet, epidote, vesuvianite, scapolite, and amphiboles such as pargasite. Ti-rich augite is found in some alkaline dyke rocks (Deer, Howie, and Zussman 1992).

Augite typically forms dark greenish black to black or brown thick prismatic crystals or massive to granular aggregates. It has been documented at over a thousand locales worldwide. Well-formed crystals of interest to collectors can be found at many places, including the following: the Yates mine, Otter Lake, Quebec, Canada; at Cedar Butte, Tillamook County, Oregon; at Mount Teide, Canary Islands, Spain; on Mount Mézenc, Haute-Loire, France; at Paškapole, near Borislav, and at several other places in the Czech Republic; and at Mount Vesuvius and Mount Etna, Italy.

Augite occurs in many meteorites, including the following: Shergotty, India; Efremova, Kazakhstan; Dar al Gani, Libya; Allende, Mexico; several presumed Martian meteorites in Morocco; Grove Mountain, Antarctica; and Lakhbi, Algeria. Its presence on Mars has been surmised from spectrometer data. Zoned augites are abundant in the Apollo 11 lunar dolerites and microgabbros, often intergrown with pigeonite, and were also found in the porphyritic lunar basalts of *Oceanus Procellarum*.

Under hydrothermal conditions, augite may be altered to chlorite and, in a manner similar to diopside it is sometimes pseudomorphously altered to amphibole in a process referred to as *uralitization*.

Figure 50. A crude black augite crystal about 1 cm long in friable matrix from Les Chazeaux, France. *RJL955*

### The Minerals: Augite   45

Figure 51. Lustrous black augite crystal group from Bancroft, Ontario, Canada; the largest crystals are about 25 mm tall. *RJL3579*

Figure 53. Group of augite single crystals from a variety of locales; largest crystal is about 2 cm tall. Top left: Zim, Czech Republic; top right: Parkapole, Petlice, Czech Republic; bottom left: Tenerife, Canary Islands, Spain; bottom right: Tanzania.

Figure 52. Dark green augite crystal about 4 cm tall, with another partial crystal, from Pontiac Co., Quebec, Canada. *RJL2834*

## Clinoenstatite

Clinoenstatite, the monoclinic dimorph of enstatite, rarely forms large crystals. It is a component of basalts near the Indigirka River, Sakha Republic, Russia, and in Uganda; and in gabbro at Lindenberg, South Africa. It is found as phenocrysts at Cape Vogel, Papua-New Guinea (regarded by some authors as the type locale).

It is a common accessory mineral in meteorites, including the following: Tenham, Australia; Indarch, Azerbaijan; Khor Temiki, Sudan; Utzenstorf, Switzerland; Leoville, Kansas; Shallowater, Texas; and Washougal, Washington.

## Clinoferrosilite

Clinoferrosilite was first reported by Bowen (1935), shortly after his own phase equilibrium studies had predicted that "pure" $Fe_2Si_2O_6$ does not exist as a crystalline compound! The mineral was discovered in the lithophysae of rhyolitic obsidian collected by Bowen near Lake Naivasha, Kenya. It formed minute (5 X 50 μm) needles that were transparent and nearly colorless. Associated minerals in the lithophysae were anorthoclase, cristobalite, magnetite, fayalite, and biotite.

Later microprobe analysis by M. G. Bown of Bowen's material indicated that the clinoferrosilite crystals contained about 5% $Mn_2Si_2O_6$ in solid solution, which possibly influenced the stability of the natural mineral compared to those used by Bowen in laboratory studies (Bown 1965).

Systematic examination of lithophysae in other obsidians found clinoferrosilite needles at three locales: In the Coso Mountains, California, the pyroxene needles were numerous but very small and were only found by using carefully chosen immersion liquids based on the earlier studies of Kenyan material. At Hrafntinnuhryggur, Iceland, the needles were very rare and also quite small. At Obsidian Cliff, Yellowstone National Park, small amounts of crystals were found that were of comparable size to the ones collected in Kenya (Bowen 1935). Thus, clinoferrosilite is probably fairly widespread in obsidians of this type, but at the same time virtually impossible to detect without advanced instruments.

The species is also found as tiny crystals in basalt at Bellerberg, near Mayen, Eifel Mountains, Germany.

Figure 54. Obsidian from Coso Hot Springs, California, containing lithophysae 2 - 3 cm in diameter. Microscopic needles of clinoferrosilite may be found in the cavities, along with glassy cristobalite and tiny brown fayalite crystals. RJL2836

# Diopside

Diopside is a widespread mineral found worldwide in many igneous and metamorphic rocks. It is particularly characteristic of contact metamorphosed Ca-rich sediments, but is also formed in regional metamorphic terrains, and is commonly found in Ca- and Mg-rich schists derived from both igneous and sedimentary rocks. Diopside is a characteristic component of many skarns where it may be associated with chondrodite or clinohumite, forsterite, magnetite, and monticellite. Diopside is the typical clinopyroxene found in ultrabasic and ultramafic igneous rocks such as peridotites and lherzolites; Cr-rich diopside is found in peridotites and nodules in kimberlite. Diopside often forms phenocrysts in basalt (although early-crystallizing pyroxene in basaltic magma tends to be augite). The mineral can occur in some alkaline igneous rocks such as jumellites, which have a distinctive mineralogy that includes Cr-rich diopside, forsterite, Ti-rich phlogopite, kataphorite, and sanidine. Al-rich diopside is the characteristic ferromagnesian component of tephrite rocks at Vesuvius and Monte Somma, Italy (Deer, Howie, and Zussman 1978).

Diopside occurs at over two thousand documented locales worldwide. Selected occurrences of particular interest to collectors will be discussed here.

The Jeffrey mine, Asbestos, Quebec, Canada, although perhaps more notable as a source of multicolored vesuvianites, has produced excellent specimens of colorless to pale green diopside, often made more desirable by colorful garnet crystals (Grice and Williams 1979; Amabili, Miglioli, and Spertini 2004). Similar examples of diopside with grossular var. *hessonite* are found at Eden Mills, Vermont. Lathlike diopside with tiny dark green grossular is found at the Orford nickel mine, Quebec, Canada.

Figure 55. Pale pink 3-mm grossular crystal on slightly curved greenish laths of diopside from the Jeffrey mine, Quebec, Canada. *RJL3326*

Figure 56. Colorful thumbnail-sized specimen of grossular var. *hessonite* on green diopside from the Jeffrey mine, Quebec, Canada. *RJL1564*

Figure 57. Colorless prismatic diopside crystals to about 5 mm on matrix from the Jeffrey mine, Quebec, Canada. *RJL2732*

Figure 58. A thumbnail-sized example of a less-common habit seen in diopside from the Jeffrey mine: thin, nearly colorless tabular crystals in random aggregates. Crystals of this habit are occasionally associated with colorful vesuvianite crystals. *RJL3407*

Diopside is known from numerous locales within the Precambrian metasedimentary rocks of the Grenville Province, but among these occurrences, De Kalb, New York, stands out for the quality of the crystals found there. The locality has been known since the late 1880s and the quality of the diopside there was mentioned by Kunz (1892). The site has been worked sporadically over the years; excellent specimens recovered in the early 1970s included superb green crystals to 8 cm tall as well as classic examples of tremolite pseudomorphs after diopside (Robinson 1990).

Figure 59. Prismatic diopside crystal about 4 cm tall, from the classic American locality in DeKalb Township, St. Lawrence County, New York. *RJL3400*

Figure 60. Group of pale green diopside crystals to about 15 mm on matrix from DeKalb, New York. *RJL3445*

Many productive localities are found throughout Europe. For example, cleft type deposits at Bellecombe, Val d'Aosta, and in Val d'Ala, both in Italy, have yielded fine crystals associated with other colorful species such as vesuvianite and grossular var. *hessonite*. *Traversellite*, an altered diopside similar to *uralite*, was described from Traversella, Piedmont, Italy.

Figure 62. Green prismatic diopside crystal, about 4 cm tall, with a complex, 2-cm black spinel crystal and massive calcite, from the Aldan Shield, Yakutia, Russia. *RJL2239*

Figure 61. Dull green diopside crystals, about 15 mm tall, from Val d'Aosta, Italy. *RJL3459*

Diopside is found in many places in India, including the Hyderabad district, and the Dumka Gem mine, Rajasthan. Star diopside is found in Nammakal, Tamilnadu, India. Cabochons showing a four-rayed star (Figure 8) have been used as a low-cost substitute for black star sapphire. Transmission electron microscopy has shown that the asterism can be attributed to microscopic inclusions of magnetite and tremolite aligned parallel to the [001] and [100] planes in the diopside crystal (Doukhan et al. 1990).

The Aldan Shield is a large, richly mineralized area in Yakutia (Sakha Republic), Siberia, Russia. Many locales in the area are notable producers of interesting diopside specimens. At Emel'dzhak and Gonovskoe, diopside is associated with lustrous, complex black spinel crystals in coarse calcite. Specimens of this material are reasonably plentiful but in most cases the accompanying label will simply identify the locale as "Aldan, Russia". Gem-quality chrome diopside found at Inagli was sometimes referred to as "inaglite", "Siberian emerald", and "Yakutian emerald". An excellent review of the mineral deposits of the Aldan Shield was given by Litsarev et al. (1997).

Figure 63. Two dark green prismatic diopside crystals in parallel growth, about 25 mm tall, from Jaipur, India. *RJL3320*

Excellent crystals are found at numerous locales in the Skardu district and the Gilgit district, Northern Areas, Pakistan. Some of this material is offered as "chrome diopside" but in many cases the green color is more likely due to iron than chromium. In Afghanistan notable occurrences include the Kunar Valley, Nuristan province, and several locales in Badakhshan and Nangarhar provinces. Among locales in China, diopside is found in Handan prefecture, Heibei province. In the Kunlun Mountains, Xinjiang, green crystals are found that are very similar to those found at DeKalb, New York.

Figure 64. Small plate of tan, 5-mm diopside crystals with grossular var. *hessonite*, from Mana, Bajaur Agency, FATA, Pakistan. Note the similarity to some of the Jeffrey mine samples. *RJL3325*

## 52 The Minerals: Diopside

Figure 65. An interesting diopside "faden" about 5 cm long, from Alchuri, Shigar Valley, Northern Areas, Pakistan. *RJL3324*

Figure 67. Thick prismatic diopside crystal about 25 mm tall, also from Turmiq, Pakistan. The color and habit are similar to material from DeKalb, New York. *RJL2469*

Figure 66. Pale green prismatic diopside crystals to about 8 mm, on matrix, from Turmiq, Skardu, Pakistan. *RJL2470*

The Minerals: Diopside 53

Figure 68. Cabinet-sized specimen of rich green "chrome diopside" on matrix from the Kunar Valley, Afghanistan. Largest crystal is about 3 cm long. *RJL2799*

Figure 69. Lustrous "chrome diopside" crystals in subparallel groups from Khugani, Nangarhar, Afghanistan. Specimen is about 4 X 6 cm. *RJL3395*

Figure 70. Dark green, partly transparent diopside crystal, about 2 cm tall, on matrix from Fargon Meeru, Badakshan, Afghanistan. *RJL3323*

54   The Minerals: Diopside

Figure 71. Diopside crystal about 25 mm tall with a somewhat unusual morphology and showing very prominent parting, also from Fargon Meeru, Afghanistan. *RJL3538*

Figure 72. Green gemmy diopside crystal about 3 cm tall, from Xinjiang, China; note the similarity to diopside from DeKalb, New York. *RJL3322*

Cr-rich varieties ("chrome diopside") are found in many places throughout the world, although in many occurrences the material has a "rock-forming" or granular habit of limited interest to collectors. Deep green chrome diopside is found at Outokumpu, Finland, where it occasionally forms large crystals. Brilliant, emerald green crystals associated with graphite and other minerals, found in the Merelani Hills, Tanzania, are often faceted.

Figure 73. Dark green chrome diopside crystal about 5 cm tall, with massive colorless quartz, from the Outokumpu mine, Finland. *RJL3406*

## 56    The Minerals: Diopside

Figure 74. Bright green chrome diopside crystal about 25 mm tall, from the Merelani Hills, Arusha, Tanzania. RJL3533

Figure 75. Sharp green chrome diopside crystal about 15 mm tall, with graphite, from the Merelani Hills, Arusha, Tanzania. RJL3321

Figure 76. Pale green diopside crystal about 3 cm tall, from the Merelani Hills, Arusha, Tanzania. This interesting floater is doubly terminated, but different forms are displayed in the terminations on the two ends. RJL3559

The term *schefferite* was applied to varieties of diopside containing Mn; at Sterling Hill, New Jersey, sharp crystals up to about 6 cm, and containing up to nearly 10% Mn, were occasionally found in limestone. *Zinc schefferite*, containing 5-7% Mn and over 3% Zn, was found abundantly in the Parker shaft at Franklin, New Jersey. This material was described as follows: "Zinc schefferite forms coarse granular or foliate masses, mixed with willemite and franklinite. A basal parting, due to lamellar twinning parallel to the base, is so perfect as to give the mineral a foliated appearance strongly suggesting feldspar..." (Palache 1935).

Figure 77. Dark brown diopside var. *schefferite* on massive calcite, from Franklin, New Jersey. As is typical of the locale, the calcite is fluorescent red under UV light. *RJL3474*

58 The Minerals: Diopside

Figure 78. Tan platy masses of diopside var. *zinc schefferite* from Franklin, New Jersey, associated with black franklinite and small off-white masses of willemite. The willemite is fluorescent bright green under UV light. Sample is about 5 cm wide. *RJL3649*

The varietal term *coccolite* was originally applied to aggregates of dark, rounded granular crystals associated with the metamorphic iron ore deposits at Arendal, Norway. Many early authors (Comstock 1859) considered this material to be a variety of augite, before the modern conception of pyroxene species and varieties had been fully developed. However, analyses have shown that the *coccolite* at Arendal is substantially a Fe-rich diopside (Nijland, Zwaan, and Touret 1998).

Figure 79. Hand specimen of diopside var. *coccolite* from Arendal, Norway. *RJL3523*

Figure 80. Detail of the specimen in the previous photo, showing the small rounded grains for which the variety is named. *RJL3523*

The varietal name *violan* or *violane* has been applied to violet-colored clinopyroxenes, which were traditionally interpreted to be Mn-rich diopside or augite. In one case, material from the Praborna manganese deposit, near St. Marcel, Italy, was studied by Mottana et al. (1979), who noted two somewhat different habits, viz., euhedral crystals in vugs and massive lamellar to fibrous aggregates. The euhedral crystals were predominantly omphacite, whereas the lamellar aggregates were mainly disordered impure diopside. The authors noted that their studies "did not clearly solve the question of the oxidation state of manganese because of the low amounts of manganese present in 'violan'." A detailed analysis of violet diopside from a calc-silicate lens in marble from Baffin Island, Canada, found *no detectable Mn*, but 0.60% Ti. The violet color in that material was attributed to intervalence charge transfer between $Fe^{2+}$ and $Ti^{4+}$ (Herd, Peterson, and Rossman 2000).

Figure 81. Rich purple diopside var. *violan* from St. Marcel, Val d'Aosta, Italy. This interesting old specimen has labels from Foote Mineral Co. as well as the original price (25 cents!) affixed to the back. The specimen is about 5 cm tall. *RJL3643*

Pseudomorphs of fibrous tremolite-actinolite after diopside are called *uralite*. This type of pseudomorph has been reported from about fifty localities worldwide. A few notable examples include: the Calumet mine, Chaffee County, Colorado; De Kalb, New York; and Traversella, Italy.

Figure 82. Actinolite replacing small prismatic diopside crystals from the Calumet mine, Chaffee County, Colorado. *RJL3146*

Figure 83. Another example of *uralite* from the Calumet mine: a cabinet-sized specimen with crystals to 15 mm long. Note the silky, nearly asbestiform nature of the actinolite replacement. RJL3473

Figure 84. Diopside var. *traversellite* from the type locale, Traversella, Italy. This older piece has passed through four collections over the years, and is an example of the same sort of replacement referred to as *uralite*. RJL2916

# Donpeacorite

Donpeacorite is a rare manganese-containing orthopyroxene described from the 2500 level of the Balmat No. 4 mine, near Balmat, New York. The mineral is found in pods in manganese-rich siliceous marbles. It forms small (1-3 mm) interlocking grains that make up over 50 percent of a massive, coarse-grained yellow-orange rock. The groundmass of this rock is mostly fibrous Mn-Mg amphibole (tirodite) with minor amounts of tourmaline, Fe-rich braunite, Mn-rich dolomite, apatite, and anhydrite. Although this mineral is very rare, and mainly of interest to the species collector, its discovery provided valuable insights into the equilibrium relations in the system $MnSiO_3$-$MgSiO_3$-$CaSiO_3$ (Peterson, Anovitz, and Essene 1984). Hypothetical end-member donpeacorite is the orthorhombic dimorph of kanoite. Both species have been reported from contact metamorphic manganese ore at Tatehira, Hokkaido Island, Japan.

Figure 85. Dark red-orange grains and masses of donpeacorite from the type locale, the Balmat No. 4 mine, near Balmat, New York. Sample is about 4 cm tall. *RJL3639*

# Enstatite

Enstatite is widely distributed throughout the world in basic and ultrabasic rocks including basalts, gabbros, peridotites, and pyroxenites. It is a characteristic component of charnockite, a coarse, granular rock that consists mainly of quartz, feldspar, and enstatite var. *hypersthene*. Enstatite is also found in lherzolite, an ultrabasic rock that consists mainly of olivine along with ortho- and clinopyroxenes. It is probably an important constituent in the upper mantle of the earth (Deer, Howie, and Zussman 1978).

Figure 86. A slab of enstatite var. *hypersthene* from Canada, showing the complex herringbone pattern resulting from an exsolution process. The exsolution lamellae will create an effect similar to tiger-eye when polished, as shown in Figure 3. *RJL3467*

Enstatite occasionally forms good crystals of interest to the collector. For the micromount enthusiast, some reported locales include the following: Puy-de-Dome, France; Emmelberg Mt., Rheinland-Palatinate, Germany; Summit Rock, Klamath Co., Oregon; and the Somma-Vesuvius Complex, Campania, Italy. Gem-quality crystals are found at Mogok, Myanmar; near Gairo, Tanzania; and at several places in Sri Lanka. Large crystals are also reported from several localities in Norway and Brazil.

Enstatite occurs in many meteorites, including: Allan Hills, Antarctica; Campo del Cielo, Argentina; Mount Egerton, Australia; Indarch, Azerbaijan; Albee, Alberta, Canada; Tagish Lake, Yukon Territory, Canada; Qingzhen, China; Bustee, India; and Gibeon, Namibia. It has also been seen in dust particles collected from the comet Wild 2 (A'Hearn 2006; Brownlee et al. 2006; Keller et al. 2006; Zolensky et al. 2006).

Figure 87. Tiny brown crystals of enstatite var. *hypersthene* from Summit Rock, Klamath Co., Oregon. Largest crystal is about 1 mm long. RJL3647

Figure 88. Gem-grade enstatite var. *hypersthene* crystal, about 35 mm tall, from Tanzania.

# Esseneite

Esseneite is a very rare pyroxene with an interesting paragenesis: it is formed through localized pyrometamorphic processes associated with the natural combustion of coal seams or coal-bearing mine dumps. It was defined as a new species based on a dominant amount of trivalent iron on **M1**, compensated by sufficient tetrahedrally coordinated aluminum. The average composition based on microprobe analysis was given as: $(Ca_{1.01}Na_{0.01})(Fe^{3+}_{0.72}Mg_{0.16}Al_{0.04}Ti_{0.03}Fe^{2+}_{0.02}Mn_{0.00})(Si_{1.19}Al_{0.81})O_6$. The simplified formula representing the pure end-member becomes: $CaFe^{3+}AlSiO_6$, a composition that had previously been synthesized but had not been found in nature. At the type locale, near Gillette, in the Powder River Basin, Wyoming, esseneite is found as small (2-8 mm) reddish-brown prismatic crystals in the slag-like paralava adjacent to a combusted coal seam. Associated phases in the paralava include anorthite, melilite solid solutions, magnetite-hercynite solid solutions, and a small amount of glass (Cosca and Peacor 1987).

Samples formed under similar circumstances in the Chelyabinsk coal basin, Southern Urals, Russia, have been described by some workers as esseneite, and a simplified formula calculated on the basis of bulk chemical analysis, $Ca_{1.01}(Fe^{3+}_{0.51}Mg_{0.45}Ti_{0.02})(Si_{1.43}Al_{0.60})O_6$, would seem to support this. However, some evidence suggests that part of the $Fe^{3+}$ is actually in tetrahedrally-coordinated sites, leading to the revised formula $Ca_{1.00}(Mg_{0.45}Fe^{3+}_{0.35}Ti_{0.02}Al_{0.18})(Si_{1.42}Al_{0.42}Fe^{3+}_{0.16})O_6$. The alert reader will notice that in the revised formula, Mg becomes the dominant occupant of **M1** rather than $Fe^{3+}$, and on that basis, the material would properly be considered Si-poor, Fe- and Al-rich diopside! The authors of that study attribute the high proportion of $Fe^{3+}$ in the tetrahedral site to the unique conditions in the Chelyabinsk occurrence, viz., high temperatures and very oxidizing conditions (Kabalov et al. 1997). However, another paper on this subject pointed out that "physicochemical conditions change drastically even on the scale of microvolumes in the heaps. Thus, there is an equal probability of finding disordered clinopyroxenes (Subsilicic ferrian aluminian diopsides) as well as ordered esseneites in the object under consideration." (Sokol, Volkova, and Lepezin 1998) For the silicate mineral collector, this situation is not uncommon: the reality on the ground is more subtle and complex than what can be easily captured on a label. As always, the collector's goal should be to *understand the specimen and what it can teach us about its environment and history.*

Figure 89. Minute amber crystals of esseneite in a chip of paralava about 2 mm across, from the type locale at the Durham Ranch, near Gillette, Wyoming. RJL3650

# Ferrosilite

Ferrosilite, the Fe-dominant end member of the enstatite-ferrosilite series of orthopyroxenes, is a characteristic component of eulysite, a type of regionally metamorphosed Fe-rich sedimentary rock. The mineral assemblage in these rocks typically includes fayalite, hedenbergite, grunerite, and garnet of the almandine-spessartine series in addition to the ferrosilite (Deer, Howie, and Zussman 1992).

Ferrosilite typically forms brown to greenish brown and black granular aggregates and occasionally prismatic crystals. It is probably fairly widespread in nature (although compositions approaching the "pure" $Fe_2Si_2O_6$ end member are rare) but collectors will usually not encounter well-documented specimens.

Reported localities include: Bear Mountain, New York; the Moxie pluton, Maine; the Loften Islands, Norway; the Kola Peninsula, Russia; in gabbro of the Guadaloupe complex, California; and in rhyolite pumice at Taupo, New Zealand.

Figure 90. Tiny, dark granular to lathlike ferrosilite crystals in matrix, from the Moxie pluton, Maine. *RJL3642*

# Hedenbergite

Hedenbergite is the Fe-dominant end member of the diopside-hedenbergite series and was first described by Berzelius in 1819 based on material from a locality in Sweden (the Tunaberg skarns, Södermanland, according to some sources). It is less common than diopside, with about 400 documented occurrences, but fine specimens from a number of locales worldwide are readily available to the collector. Hedenbergite is found in some syenite rocks such as pulaskite and foyaite, usually associated with fayalite. It is found in many metamorphic rocks, including contact-metamorphosed Ca-rich sediments and particularly contact-metamorphosed Fe-rich sediments. In regionally metamorphosed ironstones and eulysites it may be associated with fayalite and grunerite. In skarns, typical associates include garnet, bustamite, vesuvianite, scapolite, and members of the humite group (Deer, Howie, and Zussman 1992).

Hedenbergite occasionally forms short prismatic or tabular crystals, but is more commonly found as radiating to subparallel aggregates in various shades of dark green to brown or black.

Figure 92. Green, bladed hedenbergite crystals to 1 cm tall in divergent sprays, with orange-brown andradite garnet from Dal'negorsk, Russia. RJL3937

Figure 91. Fanlike aggregates of Mn-rich hedenbergite, seen in this hand-sized specimen from the Old Westinghouse property, Duquesne mine, Arizona, illustrate a typical habit of the species. RJL 3655

At present, the skarn deposit at Dal'negorsk, Russia, is perhaps the premier locality for excellent hedenbergites displaying a wide range of habits and interesting associates. It is a "principle rock-forming mineral" at the polymetallic deposits, where light to dark green or brown crystals form large radiating aggregates associated with brown spherulites of sepiolite, lustrous black ilvaite crystals, and dipyramidal quartz. It is "one of the most abundant" minerals at the borosilicate deposit, comprising a major portion of the hedenbergite-datolite and wollastonite-hedenbergite skarns (Moroshkin and Frishman 2001).

Figure 93. Thick green hedenbergite crystals with brown spherules of sepiolite, from the polymetallic deposit at Dal'negorsk, Russia. Sample is about 8 cm wide. *RJL2423*

Figure 95. An interesting plate of thin tabular hedenbergite crystals, each about 2-3 mm wide and less than 1 mm thick, associated with minor quartz, from Dal'negorsk. *RJL3069*

Figure 94. Lustrous, dull green hedenbergite from Dal'negorsk. Specimen is about 7 cm across. *RJL2784*

Figure 96. Detail of the wollastonite-hedenbergite skarn from Figure 10, showing layers of finely intergrown wollastonite (off-white) and hedenbergite (dark green to black), from the borosilicate deposit at Dal'negorsk. *RJL3531*

Fine, dark brown to black, elongated prismatic crystals are found at the Laxey mine, Owyhee County, Idaho. This deposit consists of high-grade polymetallic and replacement ores associated with an intrusive complex and its related skarn. The ore typically grades 5-10% Pb and Zn, with 7 oz/ton of Ag and significant Cu and Au values. It is probably the best locale in the United States for good hedenbergite crystals, and interestingly, excellent ilvaite crystals are also found in the district.

Figure 97. Columnar dark brown hedenbergite crystals to about 2 cm tall, with colorless quartz, from the Laxey mine, Idaho. *RJL2762*.

Lustrous black hedenbergite crystals are found at Nordmark, Sweden, associated with pyrosmalite-(Fe) and other species. That locale is notable for producing some of the world's finest pyrosmalite crystals.

Figure 98. Twinned prismatic hedenbergite crystals to about 1 cm tall, associated with brown hexagonal pyrosmalite-(Fe), from Nordmark, Sweden. *RJL2444*

Green hedenbergite-included quartz from Serifos, Greece, known since the nineteenth century, only began coming out into the general collector market in large quantities around 1988. The locality is a skarn zone created by the intrusion of granodiorite into a series of metamorphosed sedimentary strata. It includes both iron ore and Pb-Zn-Cu bodies that were mined commercially for over seventy years beginning around 1869. For collectors, the best specimens have come from vugs and fissures in the contact skarns of the marble unit. In addition to the green inclusions in quartz, hedenbergite also forms fibrous masses and dark elongated prismatic crystals associated with red-brown andradite, such as the specimen shown in Figure 29 (Gauthier and Albandakis 1991).

Other noteworthy locales include: Elba, Italy; Arendal, Aust-Agder, Norway; Dashkesan, Azerbaijan; the Skardu district, northern Areas, Pakistan; and Franklin, New Jersey. Mn-rich hedenbergite was described from a skarn at the Nakatatsu mine, Fukui Prefecture, Japan (Ito, Matsumoto, and Yoshiasa 1982).

Figure 99. Green quartz crystals about 1 cm long, colored by hedenbergite inclusions, associated with hematite "roses" from Serifos, Greece. *RJL2466*

# Jadeite

Of all the pyroxenes, jadeite is perhaps most familiar to laymen because of its longstanding importance as an ornamental stone and its significant role in even older tools and decorative artifacts dating to the neolithic period. Jadeite was especially important to some primitive societies in Central America. According to Foshag (1955), virtually all of the jade that was highly prized by the Aztecs in Mexico was jadeite (rather than *nephrite*). He points out that, "Since the oriental jadeite was unknown until its discovery in Burma in the 18th Century, A.D., the use of jadeite in America antedates its use in the Orient by more than 2000 years." Hargett (1990) presents an excellent overview of the history and rediscovery of jadeite in Guatemala, the methods by which it is mined, and the fascinating diversity of colors found there, along with a detailed comparison of the properties of Guatemalan and Burmese material.

It had been known since about 1925, based on powder diffraction data, that the structure of jadeite was closely related to that of diopside. The structure was refined by Prewitt and Burnham (1966) using a small, extremely pure crystal from the Santa Rita Peak area, California, where it is found in veinlets cutting albite-crossite-acmite schist. Compared to other pyroxenes, jadeite was traditionally thought to be fairly uncommon, but more recently it has been reported from a growing number of localities, particularly Alpine-type metamorphic settings (Deer, Howie, and Zussman 1978). Among the seventy or so documented locales worldwide, jadeite will usually be found in a dense, compact form, so with the exception of gem-quality jade, most jadeite isn't especially exciting to mineral collectors because it so rarely forms distinct crystals. However, in this regard the occurrence near Cloverdale, California, is noteworthy: here, off-white to dull gray-green jadeite forms small acicular crystals associated with serpentine, calcite, and quartz in veinlets of glaucophane rock (Wolfe 1955).

Pure end-member jadeite is colorless or off-white; shades of green are normally attributed to Fe and Cr, whereas lavender tints indicate the presence of Mn. Study of the blue-green Guatemalan jadeites indicated that the blue color component can be attributed to $Fe^{2+}$ - $Ti^{4+}$ charge transfer, whereas the green color arises from $Fe^{2+}$ and $Fe^{3+}$ absorptions (Cleary and Rohtert 2002).

Figure 100. A slice of jadeite from Myanmar (Burma) showing typically subtle variations of color from green to white. Sample is about 3 X 6 cm. *RJL172*

## 74 The Minerals: Jadeite

Figure 101. Dull green acicular jadeite crystals to about 2 mm long filling a pod in glaucophane rock from the classic locality near Cloverdale, California. Field of view is about 3 cm wide. *RJL811*

Figure 102. Samples representing various colors of Guatemalan jadeite from the Ventana Mining claim: a polished slab of New Blue, 10 X 4 cm, and (left to right) cabochons of Luna, New Blue, Princessa, Olmec Green, black, Olmec Blue, and Olmec Imperial.

## Jervisite

Jervisite, an extremely rare scandium-containing pyroxene, was described from miarolytic cavities in granite at the Diverio quarry, Baveno, Italy, where it forms tiny (200μm) pale green platelets intergrown with pink platelets of the Sc-containing pyroxenoid mineral cascandite. Microprobe analysis gave the following raw formula: $(Na_{0.43}Ca_{0.31}Mn_{0.01}Fe^{2+}_{0.29}Mg_{0.17}Sc_{0.66}Al_{0.02}Ti_{0.01})_{\Sigma=1.89}Si_2O_{6.02}$. The mineral is isostructural with the synthetic phase $NaScSi_2O_6$, and on that basis the structural formula may be expressed as: $(Na,Ca,Fe^{2+})(Sc,Mg,Fe^{2+})Si_2O_6$. Interestingly, there are only a small handful of minerals that contain Sc as an essential, major component. This is because Sc tends to be distributed in very low concentrations among the minerals that form during the early stages of magmatic differentiation, such as the pyroxenes, amphiboles, and dark mica. Thus, very little mobile Sc survives the magmatic stage to end up in pegmatites or pneumatolytic and hydrothermal rocks (Mellini et al. 1982). A review of the geology and mineralogy of the historic quarries in Baveno, Italy, was given by Albertini (1983).

Jervisite and Sc-rich aegirine have also been reported from the Deadhorse Creek complex, Ontario, Canada. Evidence indicates that it wasn't formed during magmatic differentiation, implying that Sc can be mobile under strongly alkaline conditions. The Sc "could have been introduced ... from external sources by metasomatic fluids or produced by the decomposition of previously-formed Sc-bearing phases within the main U-Be-Zr mineralization zone" (Potter and Mitchell 2005).

## Johannsenite

Johannsenite, the Mn analogue of diopside and hedenbergite, was first described by Schaller (1938) based on material from Tetela de Ocampo, Puebla, Mexico, and six other locales. This has led to some disagreement in the literature as to the type locale for the species. In fact, the IMA's Catalogue of Type Mineral Specimens lists three separate type locales: the Bohemia mining district, Lane Co., Oregon; Tetela de Ocampo, Mexico; and the Schio-Vicenza mine, near Schio, Venetia, Italy. Of these, the Mexican locale has perhaps the strongest claim, based on Schaller's statement that "...abundant material from one of these localities, namely, Puebla, Mexico, having been obtained, a more complete study of the mineral has been made. As the publication of the full report may be delayed for some time, it seems desirable to present, as an extended abstract the data upon which the new mineral is established." The type material formed greenish spherulites several cm in radius, associated with rhodonite and xonotlite. Schaller also presented analyses of johannsenites from several different locales, expressed in terms of end-member "molecules" of johannsenite, diopside, and hedenbergite, which indicated that most johannsenites tend to be close to the end-member composition ($Jo_{.80}$ to $Jo_{.95}$), although zinc-rich material from Franklin, New Jersey, had the composition $Jo_{.51}Di_{.32}He_{.09}Zn_{.08}$. Additional data from larger samples of johannsenite from Franklin were reported by Frondel (1965), which indicated a composition closer to the Mn end member.

Hutton (1956) described iron-rich johannsenite from Broken Hill, New South Wales, Australia, with an indicated composition of $Jo_{.49}He_{.46}Di_{.13}$, associated with manganpyrosmilite and bustamite. Other reported analyses of Broken Hill pyroxenes had placed them in the Fe-dominant compositional field, so those samples would properly be considered Mn-rich hedenbergite. It is likely that both species coexist at the mine.

The crystal structure of johannsenite was refined by Freed and Peacor (1967), who showed it to be isostructural with diopside. The mineral occasionally forms distinct prismatic crystals in shades of brown, gray, green, or colorless to blue. More commonly, it is found as columnar, radiating, or spherulitic masses of fibrous to prismatic crystals. It is often associated with rhodonite and with manganese oxides, which can give the crystals a dark surface.

Johannsenite is typically found in metasomatized limestones and manganiferous skarns, and, less frequently, in quartz or calcite veins. It forms solid solution series with diopside and hedenbergite, and the compositions of clinopyroxenes from skarn deposits can be expressed in terms of the proportions of Di, Jo, and He components.

In many skarns, there is a positive correlation between the He and Jo components, i.e., the Fe and Mn content of individual grains are positively correlated. Examples are known, however, in which a negative correlation is seen, i.e., as Fe increases, the Mn content decreases. Careful study of these relationships, particularly in zoned crystals, can provide insights into the types of metasomatic fluids that were involved, and how their composition fluctuated as the mineralization progressed (Nakano 1991).

Johannsenite is susceptible to various alteration processes, including oxidation, hydration, and carbonation, and it is commonly altered to rhodonite (Deer, Howie, and Zussman 1978). In a study of pyrometasomatic zinc deposits in Grant Co., New Mexico, early-crystallizing johannsenite and "ferroan johannsenite" were found to have been locally altered to rhodonite. Because the Fe-rich variety was replaced as easily as the "pure" johannsenite, it was concluded that "most of the MnO for the formation of rhodonite was introduced by the ore solutions rather than being supplied by the alteration in place of the pyroxenes with small amounts of manganese" (Allen and Fahey 1953). Johannsenite inverts to bustamite at around 830°C.

Johannsenite, with about fifty documented locales, is much less common than diopside or hedenbergite, but good crystals are known from a number of places, including: Broken Hill, New South Wales, Australia; Franklin, New Jersey; the Iron Cap mine and elsewhere in Graham Co., Arizona; the Schio-Vincenza mine, Venetia, Italy; the Bohemia district, Oregon; and the Madan district, Bulgaria.

Figure 103. Bluish green johannsenite crystals to about 1 mm long, forming rounded clusters, associated with silver, pyrite, and galena, from Broken Hill, New South Wales, Australia. RJL1271

Figure 105. Radiating johannsenite crystals forming square columns with slightly rounded terminations (similar to the habit of some Russian hedenbergites) from the Iron Cap mine, Klondyke, Graham Co., Arizona. The dark surfaces are likely due to a thin layer of manganese oxide. Specimen is about 5 cm tall. RJL2966

Figure 104. Dark green prismatic johannsenite crystals to about 5 mm, on massive sulfide-rich matrix, from Broken Hill, New South Wales, Australia. RJL2794

Figure 106. Mass of bladed johannsenite, about 5 cm tall, from the Temperino mine, Campiglia, Tuscany, Italy, one of the locales mentioned in the original description of the species (Schaller 1938). Note that some materials from this locale are identified as Mn-rich hedenbergite, although the analysis presented by Schaller was very close to end-member johannsenite. RJL3654

# Kanoite

Kanoite is a rare pyroxene described from a seam in metamorphic cummingtonite-pyroxmangite rock at the Tatehira mine, Oshima Peninsula, Hokkaido, Japan. It is found as minute (100 µm) light pinkish brown grains associated with spessartine, Mn-rich cummingtonite, and pyroxmangite (Kobayashi 1977). The species has also been reported from: Broken Hill, New South Wales, Australia; Bani Hamid, Khawr massif, Oman (Arlt and Armbruster 1997); Buritirama, Brazil; and Balmat, New York (Brown, Essene, and Peacor 1980). All of the reported occurrences are in Mn-rich amphibolite-facies metamorphic terrains (Arlt and Armbruster 1997). Kanoite is the monoclinic dimorph of donpeacorite.

Figure 107. Minute pinkish-brown masses of kanoite in a small chip of matrix from the type locale, the Tatehira mine, Oshima Peninsula, Hokkaido, Japan. Field of view is about 4 mm wide. *RJL3652*

# Kosmochlor

Kosmochlor, the $Cr^{3+}$ analogue of jadeite, has an interesting history: It was first described from the Toluca meteorite by Laspeyres (1897), who isolated 73 mg of material from 585 g of iron; he named it kosmochlor in reference to its green color and meteoritic origin. Although Laspeyres didn't determine the precise chemical makeup of the mineral, many years later Frondel and Klein (1965) described similar material from the Coahuila and Hex River Mountains meteorites and determined that all three minerals have the formula $NaCrSi_2O_6$. Frondel and Klein named their mineral *ureyite* in honor of the noted meteorite researcher Harold Urey. In the late 1980s, the name kosmochlor was reinstated by the International Mineralogical Association (Morimoto et al. 1989) based on priority of publication. Collectors should be aware that much of the literature uses the obsolete name *ureyite* (including, for example Deer, Howie and Zussman [1978]). To further complicate matters, some English-speaking authors have anglicized the name to *cosmochlore* (Couper, Hey, and Hutchison 1981; Spencer 1897).

In the Toluca meteorite, a slice of which is shown in Figure 40, tiny (0.2 mm) emerald-green kosmochlor grains occur in graphite rims of troilite nodules. Associated minerals are feldspar, zircon, quartz, and diopside containing about 1% $Cr_2O_3$. In the Coahuila and Hex River Mountains meteorites, the kosmochlor forms polycrystalline aggregates from which 200-$\mu$m cleavage fragments were isolated. These aggregates were located in nodules of daubréelite ($FeCr_2S_4$), and the kosmochlor and daubréelite were interpreted as resulting from the solidification of two immiscible liquids. In contrast to jadeite, which is only stable at high pressures, the stability field of kosmochlor extends down to ordinary pressures (Frondel and Klein 1965).

More recently, kosmochlor was recognized in the "mawsitsit" stones from serpentinized ultramafic rocks of the jade deposits of northern Myanmar. The overall appearance of the material was that of opaque, dark to bright green Burmese jade with a grain size from 0.03 to 2 mm. Samples ranged from 50 to nearly 100% kosmochlor associated with up to ~ 40% fibrous amphibole and small amounts of chromite, jadeite, and chlorite. The kosmochlor occurs in either of two forms: short prismatic crystals up to about 2 mm long, and aggregates of minute fibers (Yang 1984). More recent work has described four distinct textures involving kosmochlor and Cr-rich jadeite: 1. spheroidal or ellipsoidal aggregates having a corona texture surrounding a core of relict chromite; 2. spheroidal or ellipsoidal aggregates having a chromian jadeite core; 3. granoblastic textures in undeformed coarse-grained clinopyroxene rocks; and 4. recrystallized fine-grained aggregates in deformed jadeitite (Shi, Stockhert, and Cui 2005). The mineral has also been identified in a rock sample from Mocchi, Susa, Italy, where it forms lenses up to 3 mm wide composed of $\mu$m-sized grains in association with chromite, glaucophane, chlorite, and sphalerite (Harlow and Olds 1987). All of the foregoing terrestrial occurrences represent the products of metamorphic or metasomatic processes on ultramafic rocks.

Kosmochlor has also been found in metamorphosed Precambrian sedimentary rocks in the Slyudyanka complex, southern Baikal region, Russia. The rocks are interpreted to be metamorphic products of siliceous-carbonate sediments and include diopside quartzites to quartz-diopside rocks, diopsidites, and diopside skarns; the deposit contains sporadic zones of Cr-V mineralization and is the type locality of the V-rich pyroxene natalyite (see below). The Cr-V-rich pyroxenes are found as small (< 0.1 mm long, rarely up to ~ 0.4 mm) inclusions in quartz. The grains have a partially faceted fibrous to columnar or, more rarely, prismatic habit and are yellowish green to intense emerald green. In the same deposit, samples are also found in which Cr-rich pyroxene forms diffusion haloes around corroded eskolaite ($Cr_2O_3$) inclusions in diopside. These materials are zoned, with some areas of Cr-rich diopside and others of kosmochlor. Expressing their compositions on the basis of mol % kosmochlor, natalyite, and diopside, the kosmochlor analyses ranged from a low of $Ko_{53}Nat_{25}Di_{20}$ to a high of $Ko_{73}Nat_{18}Di_6$ (Reznitskii, Sklyarov, and Karmanov 1999).

Figure 108. A thumbnail-sized sample of *mawsitsit* from Myanmar containing dark green kosmochlore forming a tough jade-like aggregate. *RJL3532*

# Namansilite

Namansilite, the Mn analogue of aegirine, was first described from the Irnimiski deposit, Taikan Mountains, Siberia, Russia. The mineral forms dark red to red-orange grains, in veins up to 3 mm thick in braunite ores, associated with taikanite, pectolite, Mn amphibole, orthoclase, and phlogopite. Microprobe analysis gave the following formula: $[Na_{1.01}(K,Sr,Ba)_{0.01}]_{\Sigma=1.02}[Mn^{3+}_{0.93}Fe^{3+}_{0.05}Mg_{0.02}(Ti,Al)_{0.01}]_{\Sigma=1.01}Si_2O_6$. The material is isostructural with aegirine, leading to the simplified formula $NaMn^{3+}Si_2O_6$. The deposit is interpreted to be the product of the action of hydrothermal solutions related to nearby basalt intrusions (Kalinin et al. 1992).

As pointed out by Kawachi and Coombs (1993), the same mineral phase had previously been reported from a Lower Paleozoic stratiform manganese ore body at the Hoskins mine, near Grenfell, New South Wales, Australia, as well as from Val di Vara, Northern Apennines, Italy. These workers further described its occurrence at the Woods mine manganese deposit, New South Wales. All of these localities seem to have been produced by fairly low-grade metamorphism of highly oxidized, strata-bound manganiferous sediments, ultimately containing braunite and other $Mn^{3+}$ species. Deposits of this type aren't especially rare, leading the authors to note, "The fact that namansilite is so easily confused for piemontite, and that four quite independent and widely separated occurrences have been reported within a few years, suggest that this mineral may be less rare than its recent recognition might suggest."

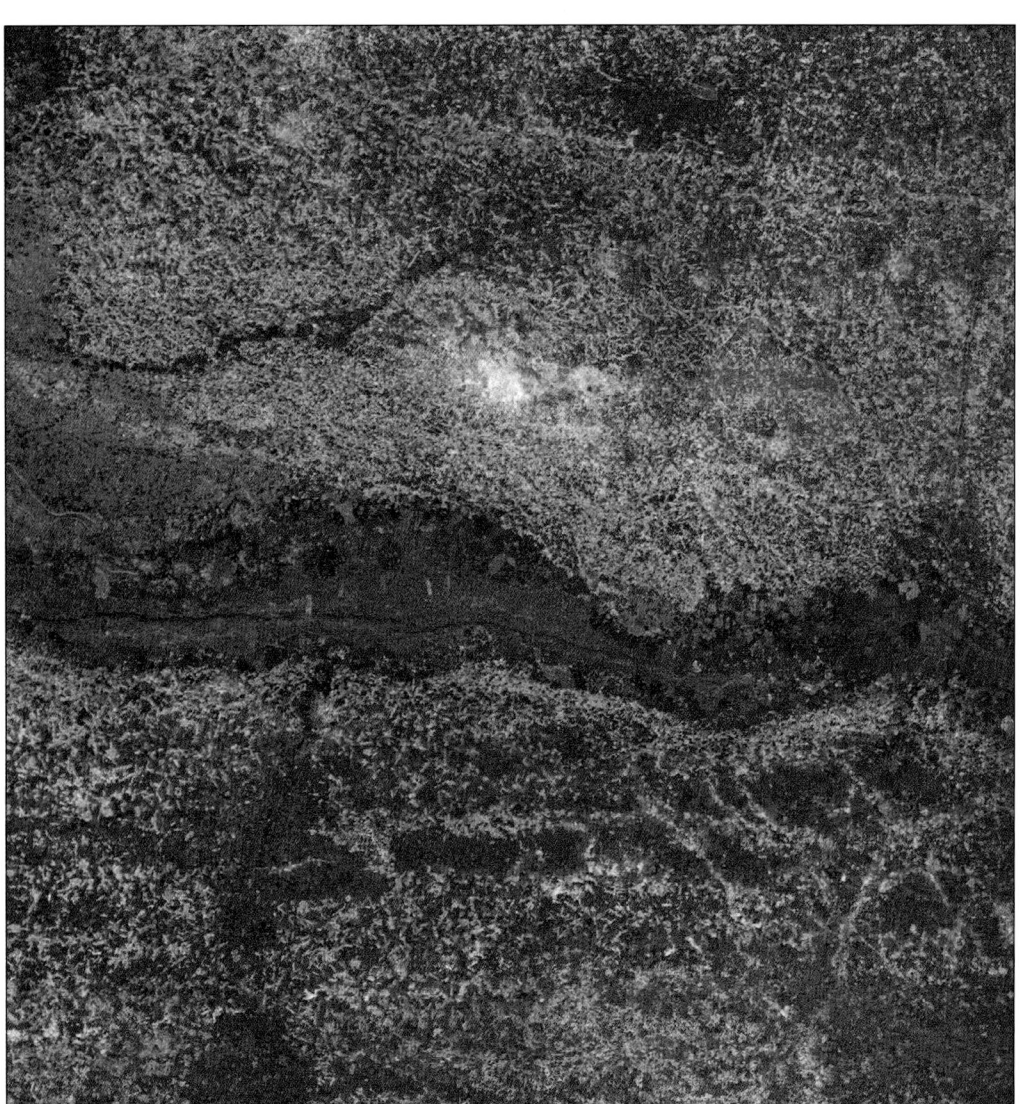

Figure 109. A vein containing red-brown namansilite, cutting through metamorphosed manganiferous sediments, from the Woods mine, New South Wales, Australia. RJL3640

## Natalyite

The chromium-vanadium pyroxene natalyite was first reported from Cr- and V-rich diopside-quartz rocks in the Slyudyanka Precambrian metamorphic complex, near Lake Baikal, Russia, where it forms small (1 X 0.3 mm) bright green grains. The mineral has a vitreous luster and a silky surface, reflecting its fibrous structure. Associated minerals include garnet (uvarovite-goldmanite series), Cr-V tourmaline, oxides of the eskolaite-karelianite series, pyrite, and apatite (Reznitskii, Sklyarov, and Ushchapovskaia 1985). It has also been reported from the Zaonezhki Peninsula, southern Karelia, Russia.

Natalyite and V-rich aegirine have been reported from the Deadhorse Creek complex, Ontario, Canada, where they form small (150 μm) lathlike crystals in quartz (Potter and Mitchell 2005).

Figure 110. Tiny dark green grains of natalyite from the type locale, the Slyudyanka Precambrian metamorphic complex, near Lake Baikal, Russia. Field of view is about 4 mm wide. *RJL3641*

## Omphacite

The name omphacite has been applied to Ca-Na pyroxenes since the early 1800s, and was generally understood to denote "green clinopyroxene occurring in eclogites and related rocks." Eclogite was defined by Hauy in 1822 as a rock with pyroxene, garnet, and kyanite as essential components. This led to a somewhat circular definition in which eclogite was defined as a rock containing omphacite and garnet, and omphacite was defined as the pyroxenitic component of eclogite (Clark and Papike 1968). Other names such as diopside-jadeite and chloromelanite have also been used to describe clinopyroxenes of this general composition. Current nomenclature recognizes omphacite as a valid species name to denote the composition range shown in Figure 15 (Morimoto et al. 1989).

Although omphacite is found in some other rock types, it is certainly best known to collectors and lapidaries as a characteristic component of eclogite, a metamorphic rock that has an overall composition much like that of basic igneous rocks, but has recrystallized under high temperature and high pressure conditions. Depending on the pressure at which eclogites form, the overall mineral assemblage changes somewhat: In eclogites that formed at very great depths, such as those of kimberlite pipes, omphacite is associated with garnet, kyanite, and corundum. In eclogites typical of migmatitic gneiss terrains, omphacite is associated with hornblende and scapolite. In eclogites of Alpine orogenic terrains, the omphacite typically has a higher content of the $NaFe^{3+}Si_2O_6$ component and is associated with a mineral assemblage that reflects lower temperatures and higher water vapor pressures, including lawsonite, pumpellyite, epidote, glaucophane, and hornblende (Deer, Howie, and Zussman 1992).

Early structural analysis of omphacite from eclogite associated with glaucophane schist from the Tiburon Peninsula, Marin County, California, suggested P2 space group symmetry (Clark and Papike 1968). Later work using a sample from the Sanbagawa metamorphic terrain near Bessi, Japan, found that this material, as well as that analyzed by Clark and Papike (1968) actually has P2/n space group symmetry (Matsumoto, Tokonami, and Morimoto 1975).

Omphacite crystals typically have a massive to granular habit, but the green omphacite and red garnets make an attractive combination in many eclogite specimens. Some noteworthy locales include the Motagua Valley, Guatemala; near Hof, Bavaria, Germany; Nové Dvory, Moravia, Czech Republic; in kimberlites at Garnet Ridge, Arizona, and in South Africa; and in glaucophane schists of the Franciscan Formation in California.

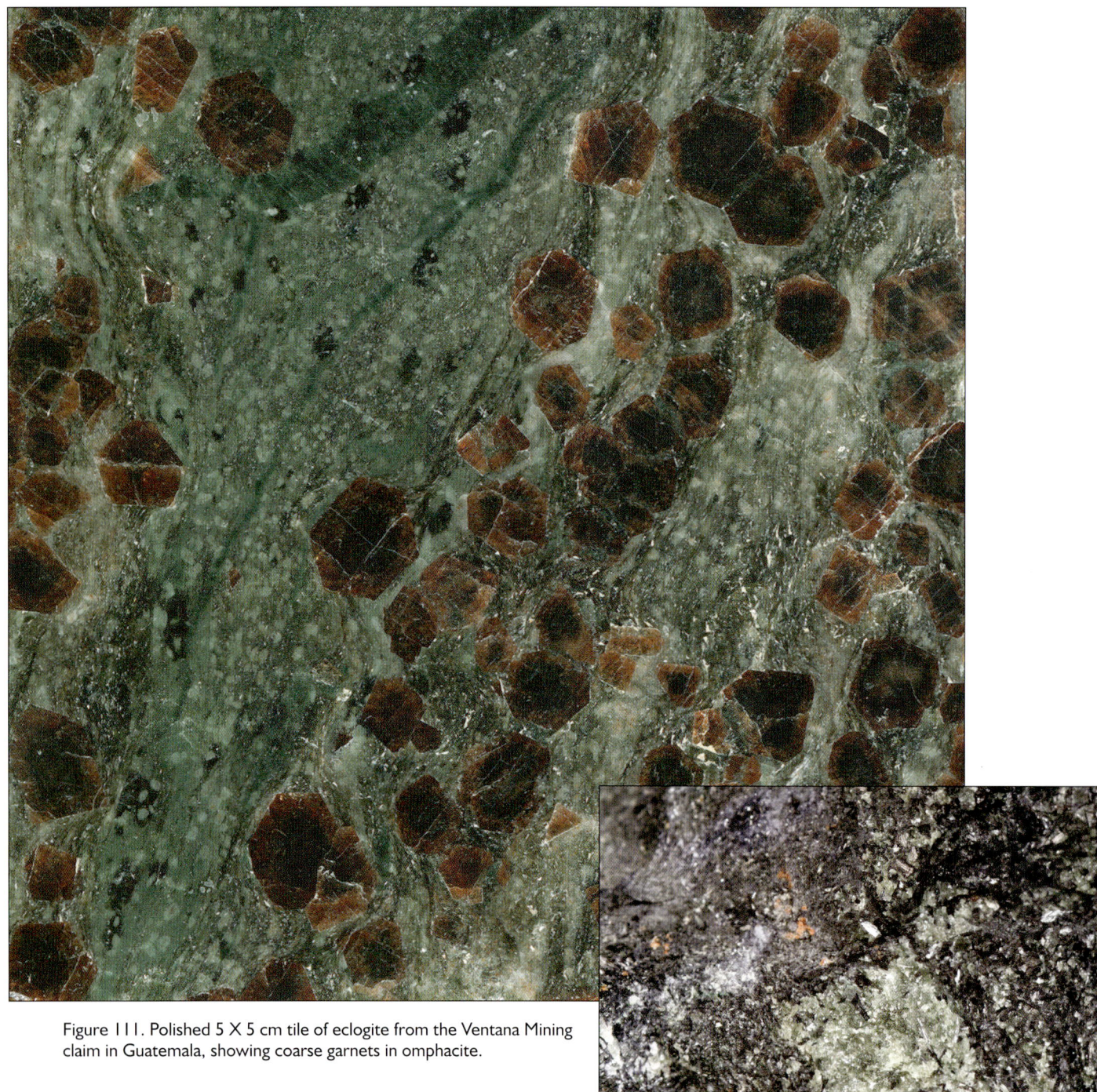

Figure 111. Polished 5 X 5 cm tile of eclogite from the Ventana Mining claim in Guatemala, showing coarse garnets in omphacite.

Figure 112. A patch of dull green omphacite about 1 cm across, in glaucophane, from Glaucophane Ridge, 5 miles north of Llanada, southern San Benito Co., California. RJL3656

# Petedunnite

Petedunnite is a rare (one-locality) Zn-dominant Ca pyroxene described from the zinc deposit at Franklin, New Jersey. Although Zn-rich clinopyroxenes such as diopside var. *jeffersonite* and diopside var. *zinc schefferite* have long been known (Palache 1935), petedunnite was the first specimen in which Zn is (just barely) dominant on the **M1** site and thus represented a valid species. The chemical formula based on microprobe analysis is: $(Ca_{0.92}Na_{0.06}Mn_{0.02})(Zn_{0.37}Mn_{0.19}Fe_{0.19}Mg_{0.14})(Si_{1.94}Al_{0.06})O_6$. The simplified end-member formula is therefore $CaZnSi_2O_6$. The original hand specimen was a 10-cm mass of dark green anhedral pyroxene surrounded by paler green pyroxene and massive calcite. In thin section, the pyroxene was found to be a mosaic of 10-100 $\mu$m subgrains, and the species was characterized by selecting only those portions with the highest Zn contents (Essene and Peacor 1987).

Since the time of its original description, petedunnite has been found in two other assemblages, the second of which was comparatively prolific. Most of the second assemblage was collected by the late John Cianciulli, curator at the Franklin Mineral Museum, who determined its optical properties; microanalysis by Tony Nikischer indicated ~ 10 to 15 weight percent Zn with low Fe. Samples from both assemblages are on display at the Franklin Mineral Museum (A. J. Nikischer, *personal communication* 2009).

Figure 113. Nondescript green masses of pyroxene, some of which is petedunnite, from Franklin, New Jersey. Analytical work has shown that it is usually the dark greenish masses that are petedunnite; when admixed in calcite, it is typically granular and paler in appearance. This sample may have both forms present. There is much solid solution with hedenbergite in the sharply crystallized, dark material, and one must always look at Zn content to determine if the mineral is present in a specific sample (A. J. Nikischer, *pers. comm.* 2009). RJL3638

# Pigeonite

Pigeonite refers to clinopyroxene compositions intermediate between those of the enstatite-ferrosilite series and those of augite, as shown schematically in Figure 13. It is a characteristic component of andesites and dacites and is found, less commonly, in basalts. The species was described by A. N. Winchell in 1900 from diabase at Pigeon Point, Minnesota. The crystal structure of pigeonite was first determined by Morimoto, Appleman, and Evans (1960) using samples collected from andesite-dacite dikes near the Asio mine, Japan.

Many pigeonites in both igneous and metamorphic rocks contain exsolution features. On cooling, the pigeonite inverts to an orthorhombic pyroxene (enstatite-ferrosilite) and the excess diopside-hedenbergite component is exsolved as thin laminae of augite generally parallel to (100) of the original pigeonite. These features have been the subject of extensive research, particularly using transmission electron microscopy, to better understand the mechanism of the exsolution process and gain insights into the thermal history of their host rocks (Robinson et al. 1971; Jaffe et al. 1975; Robinson et al. 1977).

Pigeonite forms small prismatic crystals but is often granular to massive; color ranges from pale yellow-green (in thin section) to brown and dark greenish brown. Some documented localities include: the Bushveld complex, Transvaal, South Africa; the Biwabik iron formation, Mesabi Range, Minnesota; the Black Range, Grant County, New Mexico; the Belmont quarry, Virginia; the Hakone volcano, Iwate Prefecture, Japan; the Isle of Mull, Scotland; Weiselberg, Germany; Labrador, Newfoundland, Canada; and the Skaergaard massif, Greenland.

Pigeonites occur in many achondritic or eucritic meteorites, including: Lakhbi, Algeria; Grove Mountains, Antarctica; North Haig, Australia; Ibitira, Brazil; Chassigny, France; Shergotty, India; and Dar al Gani, Libya.

Pigeonite is a common constituent of lunar basalts, and the samples retrieved during lunar missions have been studied intensively (see, e.g., Clark, Ross, and Appleman 1971).

Figure 114. Brownish prismatic to lathlike pigeonite in feldspar from the Belmont quarry, Virginia. Sample is about 3 cm wide. *RJL3648*

# Spodumene

Spodumene was originally described from Uto, Sweden, by d'Andrada (1800), who chose the name spodumene (Greek: "reduced to ashes") in reference to its typically grayish-white color. It is the only lithium pyroxene, and is a characteristic component of Li-bearing pegmatites (Deer, Howie, and Zussman 1978). In some pegmatites, notably those in Custer and Pennington Counties, South Dakota, it can form truly enormous crystals. According to Roberts and Rapp (1965): "The largest spodumene crystals known in the world have been mined from the Etta mine, Keystone. One crystal, or log, 42 feet long with a cross section of approximately 3 by 6 feet, and weighing about ninety tons, of which thirty-seven tons were commercial spodumene, has been described... In local areas the spodumene made up 50 percent of the pegmatite, and it has been estimated that it constituted 20 to 25 percent of the spodumene-bearing pegmatite." Similarly impressive crystals to about 2 meters long occur at the Harding mine, near Taos, New Mexico.

Fine, gem-quality crystals of spodumene, particularly *kunzite*, are found in the pegmatites of San Diego County, California, especially at the Himalaya and Pala Chief mines. Other notable pegmatite occurrences are: Badakshan, Kunar, Laghman, and Nargarhar provinces, Afghanistan; the Mt. Bity and Anjanabonoina, pegmatite fields, Madagascar; and numerous localities in Minas Gerais, Brazil. Emerald-green *hiddenite* containing 1200 ppm Cr is found in a mica pegmatite at Kabbur, Mysore, India. Small crystals of *hiddenite* are found sparingly at Hiddenite, North Carolina; small single crystals and (more rarely) matrix specimens are occasionally offered by mineral dealers, especially those with a local connection.

Spodumene crystals can be altered fairly easily, particularly by late-stage alkaline hydrothermal solutions. One common alteration product is a mixture of eucryptite, hexagonal $LiAlSiO_4$, and albite. The eucryptite may, in turn, be altered to Li-bearing mica; this mixture is known as *cymatolite* (Deer, Howie, and Zussman 1978).

Figure 115. A specimen with a lot of history: dull greenish spodumene from the type locale at Nykoping Gruvan, Utö Island, Mysingen Fjord, Sodermanland, Sweden, collected ca. 1850. *RJL3524*

## The Minerals: Spodumene

Figure 116. Dull green crystals of spodumene var. *hiddenite* associated with colorless quartz crystals, from Hiddenite, North Carolina. *RJL2440*

Figure 117. Green spodumene var. *hiddenite* about 1 cm tall, in pegmatite, from Hiddenite, North Carolina. *RJL3578*

Figure 118. An unusual specimen, in which several spodumene (var. *kunzite*) crystals to about 1 cm form inclusions in a water-clear beryl var. *morganite* from Mawi, Afghanistan. Note particularly the terminated spodumene in the lower right edge. *RJL31331*

88   The Minerals: Spodumene

Figure 119. Another unusual spodumene specimen: a tabular single crystal of *kunzite* about 4 X 6 cm, from Mawi, Afghanistan, containing several fluid-filled cavities with movable gas bubbles. *RJL2808*

Figure 120. Detail of the crystal in the previous photo, showing a 3-mm bubble (*see arrow*), which can move within a cavity nearly 1 cm long when the crystal is tilted back and forth. *RJL2808*

Figure 121. Sharp colorless spodumene crystal, 3 cm tall, on corroded microcline, from Dara Pech, Kunar province, Afghanistan. *RJL1596*.

# References

In addition to the references cited, readers interested in a scientific review of the pyroxene group might consider the book, *Pyroxenes*, Reviews in Mineralogy Vol. 7, C. T. Prewitt, editor, available from the Mineralogical Society of America (1980). This volume of review papers, citing over a thousand original references and reports, clearly shows the complexity of the pyroxene group and its importance to the science of petrology. The pyroxenes are exhaustively treated in Vol. 2A *Single-chain silicates*, of the Second Edition of *Rock-Forming Minerals* (Deer, Howie, and Zussman 1978).

A'Hearn, M. F. 2006. Whence comets? *Science* 314:1708-9.

Albertini, C. 1983. Famous mineral localities: Baveno, Italy. *Mineralogical Record* 14(3):157-68.

Allen, V. T., and J. J. Fahey 1953. Rhodonite, johannsenite, and ferroan johannsenite at Vanadium, New Mexico. *American Mineralogist* 38:883-90.

Amibili, M., A. Miglioli, and F. Spertini 2004. Recent discoveries at the Jeffrey mine, Asbestos, Quebec. *Mineralogical Record* 35 (2): 123-35.

Arem, J. E. 1977. *Color Encyclopedia of Gemstones*, New York: Van Nostrand Reinhold, 149 pp.

Arlt, T., and T. Armbruster 1997. The temperature-dependent $P2_1/c - C2/c$ phase transition in the clinopyroxene kanoite MnMg[$Si_2O_6$]: a single-crystal X-ray and optical study. *European Journal of Mineralogy* 9:953-64.

Bariand, P., and J. F. Poullen 1978. Famous mineral localities: The pegmatites of Laghman, Nuristan, Afghanistan. *Mineralogical Record* 9(5):301-08.

Baskerville, C., and G. F. Kunz 1904. Kunzite and its unique properties. *American Journal of Science* 18:25-28.

Bauer, M. 1904. *Precious Stones*, (1968 reprint) New York: Dover Publications, 627 pp.

Binns, R. A. 1970. Pyroxenes from non-carbonaceous chondritic meteorites. *Mineralogical Magazine* 37:649-69.

Birch, W. D., A. Wood, A. J. R. White, S. J. Mills, and R. Freeman 2008. Aegirine-augite crystals in scoria from Mt Shadwell and Mt Anakie, Victoria, Australia. *Australian Journal of Mineralogy* 14(1):37-42.

Bowen, N. L. 1935. "Ferrosilite" as a natural mineral. *American Journal of Science* 5$^{th}$ Series, 30:481-94.

Bown, M. G. 1965. Re-investigation of clino-ferrosilite from Lake Naivasha, Kenya. *Mineralogical Magazine* 34:66-70.

Brown, G. M., C. H. Emeleus, J. G. Holland, and R. Phillips 1970. Mineralogical, chemical, and petrological features of Apollo 11 rocks and their relationship to igneous processes. *Proceedings of the Apollo 11 Lunar Science Conference* Vol. 1:195-219.

Brown, P. E., E. J. Essene, and D. R. Peacor 1980. Phase relations inferred from field data for Mn pyroxenes and pyroxenoids. *Contributions to Mineralogy and Petrology* 74:417-25.

Brownlee, D., et al. 2006. Comet 81P/Wild 2 under a microscope. *Science* 314:1711-16.

Buchwald, V. 1975. *Handbook of iron meteorites: Their history, distribution, and structure*, Vol. 3, pp. 1209-10. Berkeley: University of California Press.

Cameron, M., and J. J. Papike 1981. Structural and chemical variation in pyroxenes. *American Mineralogist* 66:1-50.

Cameron, M., S. Sueno, C. T. Prewitt, and J. J. Papike 1973. High-temperature crystal chemistry of acmite, diopside, hedenbergite, jadeite, spodumene, and ureyite. *American Mineralogist* 58:594-618.

Clark, B. H., and D. R. Peacor 1992. Pyrometamorphism and partial melting of shales during combustion metamorphism: mineralogical, textural, and chemical effects. *Contributions to Mineralogy and Petrology* 112:558-68.

Clark, J. R., and J. J. Papike 1968. Crystal-chemical characterization of omphacites. *American Mineralogist* 53:840-68.

Clark, J. R., D. E. Appleman, and J. J. Papike 1969. Crystal-chemical characterization of clinopyroxenes based on eight new structure refinements. *Mineralogical Society of America Special Paper 2*: 31-50.

Clark, J. R., M. Ross, and D. E. Appleman 1971. Crystal chemistry of a lunar pigeonite. *American Mineralogist* 56:888-908.

Cleary, J. G., and W. R. Rohtert 2002. Important discovery of jadeite in Guatemala. *Gems & Gemology* 38(4):352-3.

Cleverly, W. H., E. Jarosewich, and B. Mason 1986. Camel Donga meteorite, a new eucrite from the Nullarbor Plain, Western Australia. *Meteoritics* 21(3):263-69.

Comstock, J. L. 1859. *An introduction to mineralogy.* 22$^{nd}$ Ed. New York: Pratt, Oakley & Co.

Coombs, D. S., R. J. Beck, C. J. Adams, J. M. Bannister, L. A. Patterson, and B. P. Roser 2008. Paralava produced by combustion of dead gorse near Colac Bay, Southlan, New Zealand. *Journal of Geology* 116:94-101.

Cosca, M. A., and D. R. Peacor 1987. Chemistry and structure of esseneite ($CaFe^{3+}AlSiO_6$), a new pyroxene produced by pyrometamorphism. *American Mineralogist* 72:148-56.

Couper, A. G., M. H. Hey, and R. Hutchison 1981. Cosmochlore – a new examination. *Mineralogical Magazine* 44:265-7.

d'Andrada, J. B. 1800. Der eigenschaften und kennzeichen einiger neuen fossilien aus Schweden und Norwegen nebst einigen chemischen bemerkungen ueber dieselben, *Allgemeines Journal der Chemie,* 4: 28-39, [in German; for a contemporaneous English translation, see (1801) Description of some new fossils. *A Journal of Natural Philosophy, Chemistry, and the Arts,* 5:193-196 and 5:211-213].

Deer, W. A., R. A. Howie, and J. Zussman 1978. *Rock-forming minerals, Vol. 2A Single-chain silicates,* second edition. New York: John Wiley and Sons, 668 pp.

Deer, W. A., R. A. Howie, and J. Zussman 1992. *An introduction to the rock-forming minerals,* second edition. Harlow: Pearson Prentice Hall, 696 pp.

Doukhan, N., J. Ingrin, J. C. Doukhan, and K. Latrous 1999. Coprecipitation of magnetite and amphibole in black star diopside: A TEM study. *American Mineralogist* 75:840-46.

Essene, E. J., and D. R. Peacor 1987. Petedunnite ($CaZnSi_2O_6$), a new zinc clinopyroxene from Franklin, New Jersey, and phase equilibria for zincian pyroxenes. *American Mineralogist* 72:157-66.

Finger, L. W. and Y. Ohashi 1976. The thermal expansion of diopside to 800°C and a refinement of the crystal structure at 700°C. *American Mineralogist* 61:303-10.

Foshag, W. F. 1955. Chalchihuitl – a study in jade. *American Mineralogist* 40:1062-70.

Freed, R. L. and D. R. Peacor 1967. Refinement of the crystal structure of johannsenite. *American Mineralogist* 52:709-20.

Frondel, C. 1965. Johannsenite and manganoan hortonolite from Franklin, N. J. *American Mineralogist* 50:780-2.

Frondel, C., and C. Klein 1965. Ureyite, $NaCrSi_2O_6$: A new meteoritic pyroxene. *Science* 149:742-44.

Gauthier, G., and N. Albandakis 1991. Minerals of the Seriphos skarn, Greece. *Mineralogical Record* 24(4):303-8.

Goldschmidt, V. 1922. *Atlas der Krystallformen* Vol. VII [see Facsimile Reprint in Nine Volumes (1986) by the Rochester Mineralogical Symposium].

Grant, R. and W. E. Wilson 2001. Famous mineral localities: Dal'negorsk, Primorskiy Kray, Russia. *Mineralogical Record* 32(1):3-30.

Grice, J. D., and R. Williams 1979. Famous mineral localities: the Jeffrey mine, Asbestos, Quebec, *Mineralogical Record* 10 (2): 69-80.

Grout, F. F. 1946. Acmite occurrences in the Cuyuna Range, Minnesota. *American Mineralogist* 31:125-30.

Grubb, P. L. C. 1965. An unusual occurrence of diopside and uvarovite near Thetford, Quebec. *Canadian Mineralogist* 8:241-8.

Hargett, D. 1990. Jadeite of Guatemala: A contemporary view. *Gems & Gemology* 26(2):134-41.

Harlow, G. E., and E. P. Olds 1987. Observations on terrestrial ureyite and ureyitic pyroxene. *American Mineralogist* 72:126-36.

Heffern, E. L., and D. A. Coates 2004. Geologic history of natural coal-bed fires, Powder River basin, USA. *International Journal of Coal Geology* 59:25-47.

Herd, C. D. K., R. C. Peterson, and G. R. Rossman 2000. Violet-colored diopside from southern Baffin Island, Nunavut, Canada. *Canadian Mineralogist* 38:1193-99.

Hutton, C. O. 1956. Manganpyrosmalite, bustamite, and ferroan johannsenite from Broken Hill, New South Wales, Australia. *American Mineralogist* 41:581-91.

Ito, Y., T. Matsumoto, and A. Yoshiasa 1982. A manganoan hedenbergite from the Nakatatsu mine, Fukui Prefecture, Japan and its crystal structure. *Mineralogical Journal* 11 (2): 84-92.

Ivanyuk, G. Yu. and V. N. Yakovenchuk 1997. *Minerals of the Kovdor Massif,* Apatity: RAS Kola Science Center Publishing.

Jaffe, H. W., P. Robinson, R. J. Tracy, and M. Ross 1975. Orientation of pigeonite exsolution lamellae in metamorphic augite: Correlation with composition and calculated optimal phase boundaries. *American Mineralogist* 60:9-28.

Kabalov, Yu. K., O. Oeckler, E. V. Sokolova, A. B. Mironov, and B. V. Chesnokov 1997. Subsilicic ferrian aluminian diopside from the Chelyabinsk coal basin (Southern Urals) – an unusual clinopyroxene. *European Journal of Mineralogy* 9:617-21.

Kalinin, V. V., I. M. Marsii, Yu. P. Dikov, N. V. Troneva, and N. V. Trubkin 1992. Namansilite NaMn$^{3+}$Si$_2$O$_6$: A new silicate. *Zapiski Vses. Mineral.Obshch.* 121 (1):89-94 [in Russian; see abstract (1993) *American Mineralogist* 78:1314-19].

Kawachi, Y., and D. S. Coombs 1993. Namansilite, NaMn$^{3+}$Si$_2$O$_6$: A widespread clinopyroxene? *Mineralogical Magazine* 57:533-38.

Keller, L. P., et al. 2006. Infrared spectroscopy of Comet 81P/Wild 2 samples returned by Stardust. *Science* 314:1728-31.

Kilpady, S. 1960. An X-ray study and re-examination of blanfordite. *Proc. Nat. Inst. Sci. India* 26:250-9.

Kobayashi, H. 1977. Kanoite, (Mn$^{2+}$,Mg)$_2$Si$_2$O$_6$, a new clinopyroxene in the metamorphic rock from Tatehira, Oshima Peninsula, Hokkaido, Japan. *J. Geol. Soc. Japan* 83:537-42 [see abstract (1978) *American Mineralogist* 63:598].

Kunz, G. F. 1892. *Gems and Precious Stones of North America*, Second Edition, (1968 reprint) New York: Dover Publications, 367 pp.

Laspeyres, A. 1897. Mittheilungen aus dem mineralogischen Museum der Universitat Bonn. VIII. Theil. *Zeit. Krist.* 27:586-600.

Leavitt, D. L. 1981. Minerals of the Yates uranium mine, Pontiac County, Quebec. *Mineralogical Record* 12(6):359-63.

Levien, L. and C. T. Prewitt 1981. High-pressure structural study of diopside. *American Mineralogist* 66:315-23.

Lisitsyn, A. E., and S. V. Malinko 1994. The Dal'negorsk boron deposit: a unique mineralogical object. *World of Stones* 4:30-40.

Litsarev, M. A., V. Ya. Gerasimenko, P. S. Mochalov, and E. Sh. Kudaev 1997. Minerals of the Aldan Shield. *World of Stones* 12:42-9.

Mandarino, J. A. and V. Anderson 1989. *Monteregian treasures: The minerals of Mont Saint-Hilaire, Quebec*, Cambridge: Cambridge University Press.

Marks, M. A. W., J. Schilling, I. M. Coulson, T. Wenzel, and G. Markl 2008. The alkaline-peralkaline Tamazeght complex, High Atlas Mountains, Morocco: mineral chemistry and petrological constraints for derivation from a compositionally heterogeneous mantle source. *Journal of Petrology* 49(6):1097-1131.

Mason, B. 1968. Pyroxenes in meteorites. *Lithos* 1:1-11.

Matsumoto, T., M. Tokonami, and N. Morimoto 1975. The crystal structure of omphacite. *American Mineralogist* 60:634-41.

McCoy, T. J., M. Wadhwa, and K. Keil 1999. New lithologies in the Zagami meteorite: evidence for fractional crystallization of a single magma unit on Mars. *Geochimica et Cosmochimica Acta* 63:1249-62.

McSween, H. Y. 1985. SNC meteorites: clues to Martian petrologic evolution? *Reviews of Geophysics* 23(4):391-416.

McSween, H. Y. 1987. *Meteorites and their parent planets.* Cambridge: Cambridge University Press. 237 pp.

Mellini, M., S. Merlino, P. Orlandi, and R. Rinaldi 1982. Cascandite and jervisite, two new scandium silicates from Baveno, Italy. *American Mineralogist* 67:599-603.

Mills, S. J. and L. A. Groat 2008. The crystal structure of yellow aegirine-augite from Mount Anakie, Victoria. *Australian Journal of Mineralogy* 14(1):43-5.

Mittlefehldt, D. W. 2005. Ibitira: A basaltic achondrite from a distinct parent asteroid and implications for the Dawn mission. *Meteoritics and Planetary Science* 40(5):665-77.

Morimoto, N., D. E. Appleman, and H. T. Evans, Jr. 1960. The crystal structures of clinoenstatite and pigeonite. *Zeitschrift für Kristallographie* 114:120-47.

Morimoto, N., J. Fabries, A. K. Ferguson, I. V. Ginsburg, M. Ross, F. A. Seifert, J. Zussman, K. Aoki, and G. Gottardi 1989. Nomenclature of pyroxenes. *Canadian Mineralogist* 27:143-56.

Moroshkin, V. V., and N. I. Frishman 2001. Dal'negorsk: Notes on mineralogy, *Mineralogical Almanac,* Vol. 4. Moscow: Ocean Pictures, Ltd.

Mottana, A., G. Rossi, A. Kracher, and G. Kraut 1979. Violan revisited: Mn-bearing omphacite and diopside. *Tschermaks Mineral. Petrol. Mitt.* 26:187-201 [see abstract (1980) *American Mineralogist* 65:813].

Nakano, T. 1991. An antipathetic relation between the hedenbergite and johannsenite components in skarn clinopyroxene from the Nagata tungsten deposit, central Japan. *Canadian Mineralogist* 29:427-34.

Nayak, B. R., B. K. Mohapatra, and R. K. Sahoo 1997. Mn-pyroxenes from Gangpur Group of rocks, Goriajhar, Orissa, India. *J. Min. Petr. Econ. Geol.* 92:425-30.

Nijland, T. G., J. C. Zwaan, and L. Touret 1998. Topographical mineralogy of the Bamble sector, south Norway. *Scripta Geologica* 118:1-46.

Oberti, R. and F. A. Caporuscio 1991. Crystal chemistry of clinopyroxenes from mantle eclogites: A study of the key role of the M2 site population by means of crystal-structure refinement. *American Mineralogist* 76:1141-52.

Palache, C. 1935. *The Minerals of Franklin and Sterling Hill, Sussex County, New Jersey*, Geological Survey Professional Paper 180, U.S. Government Printing Office, 135 pp.

Papike, J. J. 1980. Pyroxene mineralogy of the moon and meteorites. *Reviews in Mineralogy* 7:495-525.

Petersen, E. U., L. M. Anovitz, and E. J. Essene 1984. Donpeacorite, $(Mn,Mg)MgSi_2O_6$, a new orthopyroxene and its proposed phase relations in the system $MnSiO_3$-$MgSiO_3$-$FeSiO_3$. *American Mineralogist* 69:472-80.

Poldervaart, A. 1947. The relationship of orthopyroxene to pigeonite. *Mineralogical Magazine* 28:164-72.

Polkanov, A. A. 1939. On the gigantic aegirine-augite crystals from the plutonic rocks of Gremiakha-Vyrmes (Kola Peninsula). *Dokl. Acad. Nauk SSSR* 24:935-7.

Potter, E. G., and R. H. Mitchell 2005. Mineralogy of the Deadhorse Creek volcaniclastic breccia complex, northwestern Ontario, Canada. *Contributions to Mineralogy and Petrology,* 150: 212-229.

Prencipe, M., M. Tribaudino, A. Pavese, A. Hoser, and M. Reehuis 2000. A single-crystal neutron diffraction investigation of diopside at 10 K. *Canadian Mineralogist* 38:183-89.

Prewitt, C. T. 1980. *Pyroxenes.* Reviews in Mineralogy, Vol. 7. Washington: Mineralogical Society of America. 525 pp.

Prewitt, C. T., and C. W. Burnham 1966. The structure of jadeite, $NaAlSi_2O_6$. *American Mineralogist* 51:956-75.

Prewitt, C. T., and D. R. Peacor 1964. Crystal chemistry of the pyroxenes and pyroxenoids. *American Mineralogist* 49:1527-42.

Raudsepp, M., F. C. Hawthorne, and A. C. Turnock 1990. Evaluation of the Reitveld method for the characterization of fine-grained products of mineral synthesis: The diopside-hedenbergite join. *Canadian Mineralogist* 28:93-109.

Reznitskii, L. Z., E. V. Sklyarov, and Z. F. Ushchapovskaia 1985. Natalyite $Na(V,Cr)Si_2O_6$ – A new chromium-vanadium pyroxene from Slyudianka. *Zapiski Vses. Mineralog. Obshch.* 114:630-35 [see abstract (1987) *American Mineralogist* 72:222-230].

Reznitskii, L. Z., E. V. Sklyarov, and N. S. Karmanov 1999. The first find of cosmochlore (ureyite) in metasedimentary rocks. *Doklady Earth Sciences* 364(1):64-7.

Roberts, W. L., and G. Rapp, Jr. 1965. *Mineralogy of the Black Hills*. Bulletin No. 18 of the South Dakota School of Mines and Technology, Rapid City, South Dakota.

Robinson, G. W. 1990. Famous mineral localities: De Kalb, New York. *Mineralogical Record* 26(6):535-41.

Robinson, P., H. W. Jaffe, M. Ross, and C. Klein, Jr. 1971. Orientation of exsolution lamellae in clinopyroxenes and clinoamphiboles: consideration of optimal phase boundaries. *American Mineralogist* 56:909-39.

Robinson, P., M. Ross, G. L. Nord, Jr., J. R. Smyth, and H. W. Jaffee 1977. Exsolution lamellae in augite and pigeonite: fossil indicators of lattice parameters at high temperature and pressure. *American Mineralogist* 62:857-73.

Rogova, V. P., Yu. G. Rogov, V. V. Drits, and N. N. Kuznetsova 1978. Charoite, a new mineral, and a new jewelry stone. *Zapiski Vses. Mineralog. Obshch.* 107:94-100 [see abstract (1978) *American Mineralogist* 63:1282].

Roy, S. 1970. Manganese-bearing silicate minerals from metamorphosed manganese formations of India. I. Juddite. *Mineralogical Magazine* 37:708-16.

Roy, S. 1971. Manganese-bearing silicate minerals from metamorphosed manganese formations of India. II. Blanfordite, manganoan diopside, and broan manganiferous pyroxene. *Mineralogical Magazine* 38:32-42.

Rubin, A. E., G. J. Taylor, K. Keil, and G. Nelson 1981. The Correo and Suwanee Spring meteorites: two new ordinary chondrite finds. *Meteoritics* 16(1):9-12.

Schaller, W. T. 1938. Johannsenite, a new manganese pyroxene. *American Mineralogist* 23:575-82.

Shi, G. H., B. Stöckhert, and W. Y. Cui 2005. Kosmochlor and chromian jadeite aggregates from the Myanmar jaditite area. *Mineralogical Magazine* 69:1059-75.

Skogby, H., D. R. Bell, and G. R. Rossman 1990. Hydroxide in pyroxene: Variations in the natural environment. *American Mineralogist* 75:764-74.

Smith, J. V., D. A. Stephenson, R. A. Howie, and M. H. Hey 1969. Relations between cell dimensions, chemical composition, and site preference of orthopyroxene. *Mineralogical Magazine* 37:90-114.

Smyth, J. R., D. R. Bell, and G. R. Rossman 1991. Incorporation of hydroxyl in upper-mantle pyroxenes. *Nature* 351:732-4.

Sokol, E., N. Volkova, and G. Lepezin 1998. Mineralogy of pyrometamorphic rocks associated with naturally burned coal-bearing spoil heaps of the Chelyabinsk coal basin, Russia. *European Journal of Mineralogy* 10:1003-14.

Spencer, L. J. 1897. A list of new mineral names. *Mineralogical Magazine* 11:323-37.

Washington, H. S., and H. E. Merwin 1927. The acmitic pyroxenes. *American Mineralogist* 12:233-52.

Williams, R. J., and J. J. Jadwick 1980. *Handbook of lunar materials*. NASA Reference Publication 1057.

Wolfe, C. W. 1955. Crystallography of jadeite crystals from near Cloverdale, California. *American Mineralogist* 40:248-60.

Wood, C. A., and L. D. Ashwal 1981. SNC meteorites: Igneous rocks from Mars? *Proceedings Lunar and Planetary Sciences* 12B:1359-75.

Yamaguchi, A., R. N. Clayton, T. K. Mayeda, M. Ebihara, Y. Oura, Y. N. Miura, H. Haramura, K. Misawa, H. Kojima, and K. Nagao 2002. A new source of basaltic meteorites inferred from Northwest Africa 011. *Science* 296:334-6.

Yang, C. M. O. 1984. A terrestrial source of ureyite. *American Mineralogist* 69:1180-83.

Zolensky, M. E., et al. 2006. Mineralogy and petrology of Comet 81P/Wild 2 nucleus samples. *Science* 314:1735-39.

# More Schiffer Earth Science Monograph Titles

www.schifferbooks.com

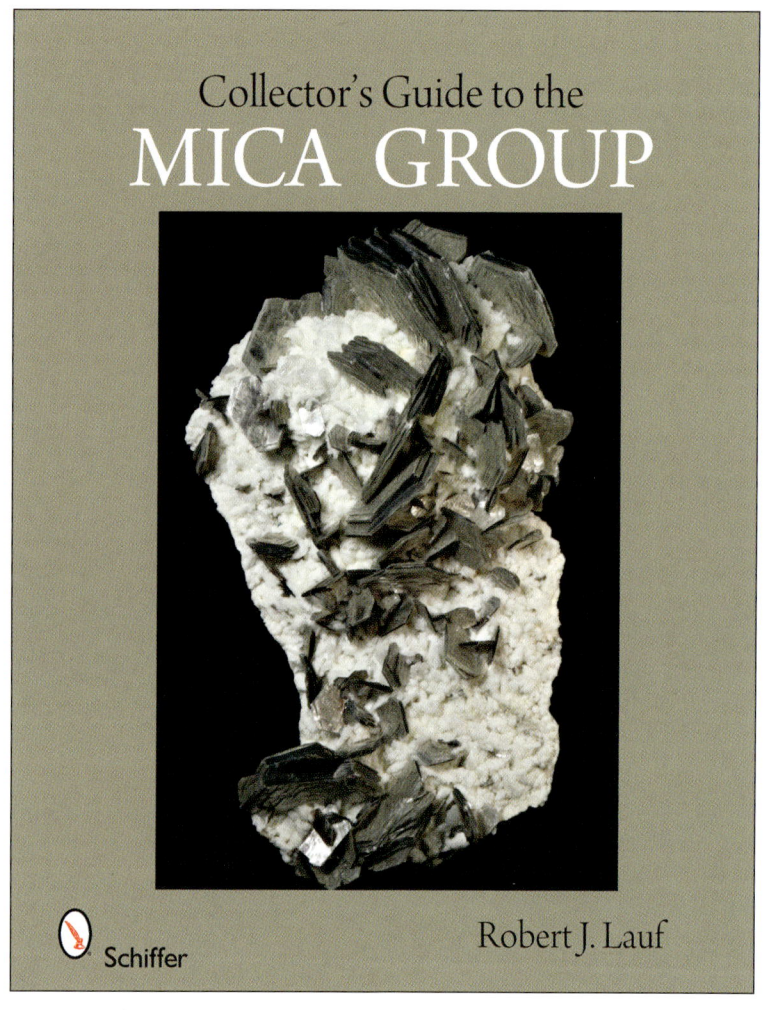

**Collector's Guide to the Mica Group.** Robert J. Lauf. Mica is a broad term encompassing about forty minerals, ranging from the common to the rare, many at times forming excellent crystals jewelers use. This book features examples recently described among the 115 striking color photos and electron micrographs that illustrate the text. A detailed entry for each type includes information on where each is found, associations of micas with other minerals, pseudomorphs (minerals that masquerade as mica), and micas that fluoresce under UV light. This fascinating guide is for those interested in minerals.

| | | |
|---|---|---|
| Size: 8 1/2" x 11" | 115 color photos | 96pp. |
| ISBN: 978-0-7643-3047-6 | soft cover | $19.99 |

# OTHER SCHIFFER TITLES

## www.schifferbooks.com

**Introduction to Radioactive Minerals.** Robert J. Lauf. Collectors have long admired uranium and thorium minerals for their brilliant colors, intense ultraviolet fluorescence, and rich variety of habits and associates. Radioactive minerals are also critically important as our source of nuclear energy. Understanding them is crucial to the safe disposal of radioactive waste. This book provides a systematic overview of the mineralogy of uranium and thorium, generously illustrated with nearly 200 color photos and electron micrographs of representative specimens. Includes an historical discussion of the discovery of radioactive elements and the development of uranium and thorium ore deposits, a discussion of the geochemical conditions that produce significant deposits, and a description of important localities, their geological setting and history. Major occurrences of interest to mineral collectors are arranged geographically. The minerals are arranged systematically, to emphasize how they fit into chemical groups, and for each group a few minerals are selected to illustrate their formation and general characteristics. With the resurgence of interest in nuclear power, this book is an invaluable guide for mineral collectors as well as nuclear scientists and engineers interested in radioactive deposits.

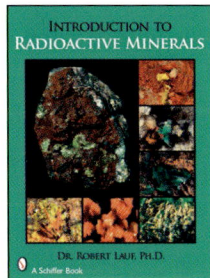

Size: 8 1/2" x 11"   196 color & b/w photos   144pp.
ISBN: 978-0-7643-2912-8   soft cover   $29.95

**Collecting Fluorescent Minerals.** Stuart Schneider. Seeing fluorescent minerals up close for the first time is an exciting experience. The colors are so pure and the glow is so seemingly unnatural, that it is hard to believe they are natural rocks. Hundreds of glowing minerals are shown, including Aragonite, Celestine, Feldspar, Microcline, Picropharmacolite, Quartz, Spinel, Smithsonite, plus many more. But don't let the hard-to-pronounce names keep you away. Over 800 beautiful color photographs illustrate how fluorescent minerals look under the UV light and in daylight, making this an invaluable field guide. Included are values, a comprehensive resources section, plus helpful advice on caring for, collecting, and displaying minerals. The field of collecting fluorescent minerals is relatively new and this is one of the most complete references available.

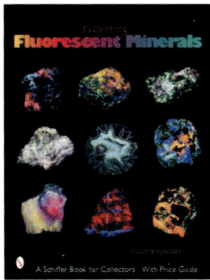

Size: 8 1/2" x 11"   846 color photos   192pp.
ISBN: 0-7643-2091-2   soft cover   $29.95

**The World of Fluorescent Minerals.** Stuart Schneider. The rich and diverse world of fluorescent minerals is explored in this sweeping survey. Breathtakingly pure colors, with their ethereal glow, immediately capture your attention. Did you know that color television is a result of the study of fluorescing minerals? Fresh finds of fluorescent minerals are showing up regularly around the globe, and their collection is an entertaining and popular past-time. To help the collector, over 825 photos display the minerals both as they might be found in daylight and in under the effects of ultraviolet light. Written for the collector and the merely curious, this pictorial reference will enrich your collecting experience with its informative text. It is an essential source for enjoying and identifying fluorescent minerals.

Size: 8 1/2" x 11"   825 color photos   176pp.
ISBN: 0-7643-2544-2   soft cover   $29.95

---

Schiffer books may be ordered from your local bookstore, or they may be ordered directly from the publisher by writing to:

Schiffer Publishing, Ltd.
4880 Lower Valley Rd.
Atglen, PA 19310
(610) 593-1777; Fax (610) 593-2002
E-mail: Info@schifferbooks.com

Please visit our web site catalog at *www.schifferbooks.com* or write for a free catalog. Please include $5.00 for shipping and handling for the first two books and $2.00 for each additional book. Full-price orders over $150 are shipped free in the U.S.

Printed in China

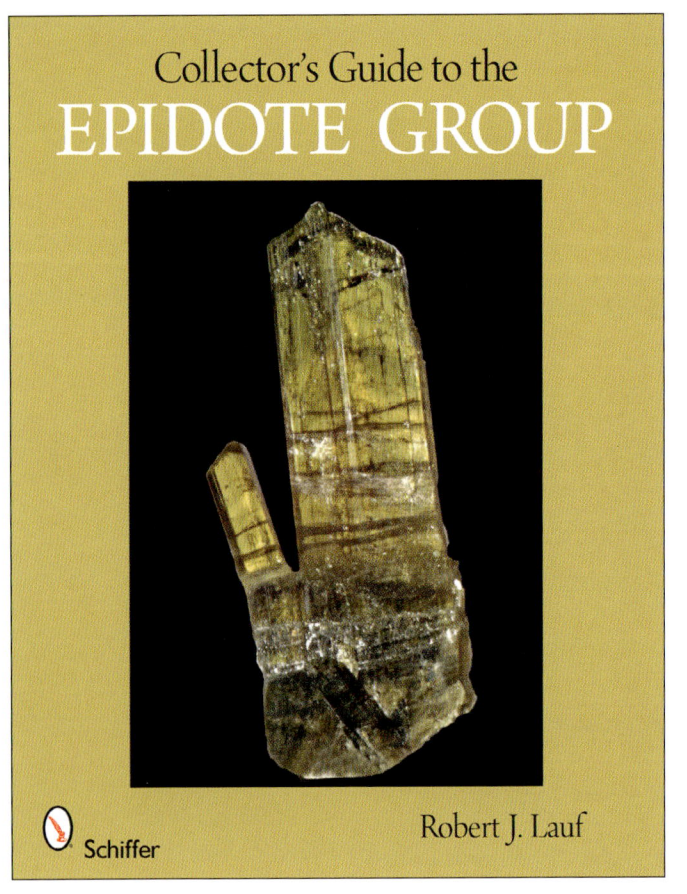

**Collector's Guide to the Epidote Group.** Robert J. Lauf. Over 90 striking color photos display minerals of the epidote group, well known to mineral collectors for their rich colors and the many interesting minerals with which they occur. Lapidary artists also value epidote, particularly in the form of unakite, and precious or semiprecious varieties of the related mineral zoisite, including thulite and tanzanite, some which have inclusions that allow them to be cut into popular catseyes. This informative book provides all presently known species, detailed entries for each of the eighteen minerals, and extensive locality information. This book will be of interest to those interested in developing a better understanding of silicate minerals.

Size: 8 1/2" x 11"  92 color photos  96pp.
ISBN: 978-0-7643-3048-3  soft cover  $19.99

**Collector's Guide to Fluorite.** Arvid Eric Pasto. Fluorite is found everywhere, has been important to industry for centuries, and is a minor ornamental material as well. Fluorite presents a fascinating array of colors, habits, and associated minerals and is widely available. See spectacularly large "museum quality" specimens that can be found. Fluorite presents a wealth of scientific opportunities to see crystallography, geochemistry, and solid-state physics at work in the natural world. It provides over 140 full-color examples and extensive references to the formation and geographic locations of fluorite. This book is essential for everyone with a passion for minerals.

Size: 8 1/2" x 11"  143 color photos, 10 illus.  96pp.
ISBN: 978-0-7643-3193-0  soft cover  $19.99